Praise for *The Education of Eva Moskowitz*

"Success Academy, run by Eva Mosk
education-policy analyst considered remotely possible: its schools have closed the achievement gap."

—Jonathan Chait, *New York* magazine

"A lively new memoir," by "one of the country's most influential crusaders at a turning point for charter schooling."

—Elizabeth Green, *The Atlantic*

"Readers get the sense that Moskowitz, fiercely independent and confrontational, likes the fight: she knows she's in it for the right reasons—building better schools—and isn't giving up."

—*Weekly Standard*

"Eva Moskowitz has become a punching bag for the Left, and most of what's said against her is plain false. While we disagree on some big things, she is a force-of-nature changemaker who has improved the lives of thousands of children, and this compelling, often amusing memoir drops the mic on her critics."

—Jonathan Alter, MSNBC contributor and author of *The Center Holds: Obama and His Enemies*

"Much has been written about Eva Moskowitz, a pioneering and controversial figure in education circles. Now she tells her own story. Whether you're a fan, a critic, or you just care about urban education, you'll be sure to learn something from this book."

—James E. Ryan, president, University of Virginia

"Education reform is the civil rights battle of our time, and this book is critical to understanding how to defeat a corrupt status quo that continues to stand in the way of innovation and improvement."

—Joe Scarborough, cohost of MSNBC's *Morning Joe*

"[Moskowitz] writes beautifully, salting her own clear, passionate, yet often analytic prose with quotes and references plucked from many sources, from Twain and Dickens to Carnegie, Fielding, and Hardy."

—*EducationNext*

THE EDUCATION *of* EVA MOSKOWITZ

ALSO BY EVA MOSKOWITZ

In Therapy We Trust: America's Obsession with Self-Fulfillment

Mission Possible: How the Secrets of the Success Academies Can Work in Any School (with Arin Lavinia)

THE EDUCATION *of* EVA MOSKOWITZ

A Memoir

EVA MOSKOWITZ

HARPER

NEW YORK • LONDON • TORONTO • SYDNEY

HARPER

A hardcover edition of this book was published in 2017 by HarperCollins Publishers.

HarperCollins books may be purchased for educational, business, or sales promotional use. For information, please email the Special Markets Department at SPsales@harpercollins.com.

FIRST HARPER PAPERBACKS EDITION PUBLISHED 2018.

Designed by Leah Carlson-Stanisic

Library of Congress Cataloging-in-Publication Data has been applied for.

ISBN 978-0-06-244979-5 (pbk)

HB 08.07.2023

It is said that truth passes through three stages. First, it is ridiculed. Second, it is violently opposed. Third, it is accepted as being self-evident.

This is a story about the second stage.

CONTENTS

THE EDUCATION *of* EVA MOSKOWITZ

1

LE BILBOQUET

2004–2005

Trouble was brewing. Ron Perelman, a financial titan, prominent philanthropist, and alumnus of several messy public divorces was opposing a sidewalk café permit application by Le Bilboquet, a chic restaurant in the district I represented on the New York City Council. This dispute had all the makings of a juicy press story and I wanted no part of it. I'd be charged with doing favors for a wealthy constituent if I opposed the application, and, if I supported it, I'd be charged with ignoring my constituents' concerns. But alas, neutrality wasn't an option since I was legally part of the approval process, which went like this: the community board held hearings and made a recommendation to the local council member (me), who made a recommendation to the council's zoning and franchising subcommittee, which held more hearings and made a recommendation to the Land Use Committee, which made a recommendation to the council, which then voted yea or nay. It took, on average, 465 days. Yes, all for a sidewalk café permit. Welcome to my world.

Now, to some, sidewalk cafés are a charming feature of city life that enliven the streets and provide opportunities to dine alfresco in the urban landscape. To others, however, they are an appropriation of public space by rapacious businessmen who obstruct already congested sidewalks and inflict noisy late-night revelry on residential neighborhoods. My constituents tended toward the latter view and could afford to care about things like this since they were a virtual who's who in the worlds of business (Michael Bloomberg, Jamie Dimon), movies (Martin Scorsese, Woody Allen, Spike Lee, Bill Murray), music (Mariah Carey), politics (Rudy Giuliani, Eliot Spitzer), art (Jeff Koons), and comedy (Joan Rivers). I represented Manhattan's famous "Silk Stocking District," which included the most expensive real estate per square foot in the country, world-famous museums, and startlingly expensive private schools with names like Spence, Chapin, and Nightingale-Bamford.

My view on sidewalk cafés was simple: I did whatever my community board told me to. They held hearings at which they heard directly from neighborhood residents—why second-guess them? Besides, I had my hands full running the committee that oversaw the city's enormous public school system. So if anybody came to see me about a sidewalk café permit, I told 'em: talk to the community board; I do whatever they tell me. And boy was I glad to have this policy now.

But one other thing worried me. Not being born yesterday, I knew my campaign staff would soon receive one of those oh-so-rare but gratifying unsolicited contributions from some good-hearted citizen who had spontaneously recognized my merits as a public servant. Then, only after depositing the check, would we learn the contributor was a lobbyist for Perelman. We could refund the contribution, but if the press had gotten wind of this story by that point, it would look like I'd done so only because I'd been caught with my hand in the cookie jar. I therefore directed my staff to scrutinize every unsolicited contribution we received and, if it was in any way associated with Perelman—even if it was from Perelman's second cousin's lawyer's podiatrist's stepbrother's mistress—to return it immediately. Sure enough, we soon got a nice fat check from a lawyer who, when we asked what had inspired his newfound generosity, disclosed that he worked for Perelman. We immediately returned the check, and when I got a phone call from Perelman I said—*oh how I loved saying!*—talk to the community board, I just work here.

So he did. In fact, he provided a video showing that Le Bilboquet was already serving drinks outside, which people would imbibe while sitting on the steps of his town house, leaving behind a trail of cigarette butts. This was a huge no-no, so the community board turned down the application. Made sense to me—but again, not my department. I passed the recommendation along to the Zoning and Franchising Subcommittee.

Just as I'd feared, the press took an interest. Not even the *New York Times,* which usually turned up its nose at boldface names' stories, could resist. Ron Perelman! A chic Upper East Side café! Yes, it was gossip, but it was the caviar of gossip. I told the *Times* reporter exactly what had happened: how I'd supported my community board's recommendation, which was always my practice, how I'd instructed my staff from the outset to return any checks they received from Perelman, and how they'd done so. I'd predicted exactly what would happen and had acted scrupulously and consistently. There was no way I could be criticized.

Oh, how naive I was.

Here's an excerpt from the article the *Times* published titled "Steak Frites and Stardom vs. Power and Politics":

> This is a story about four tables and eight chairs, the billionaire cosmetics executive who would not brook them, and an ambitious councilwoman who took a phone call. It will not end prettily . . .
>
> Mr. Perelman's lawyer . . . made a campaign contribution to Ms. Moskowitz, who wants to run for Manhattan borough president. On Tuesday, the councilwoman instructed her staff to return the check.
>
> Ms. Moskowitz said . . . she simply wanted to go along with the recommendation of the community board . . . and was too busy with issues like teachers' contracts and school construction budgets to look into the matter more deeply . . .
>
> Philippe Delgrange, the owner of Le Bilboquet . . . who is of French and Belgian descent, said the dispute feels personal . . . "Maybe he doesn't like the Belgians or French."

By saying we'd returned the check "on Tuesday," rather than "immediately," the reporter implied we'd delayed doing so until she'd come sniffing around. Moreover, she didn't mention that I always deferred to the community board or that I did so because the board heard directly from the community, so it just sounded like I was making this up or was too busy to do my job. It was a textbook example of writing the lead on the way to the ballpark.

Rule number one of journalism, I was learning, is that trying to get in between a journalist and a story he wants to tell is like trying to stop a herd of stampeding cattle. Stories headlined "Council Member Acts Ethically, Follows Policies Consistently" don't sell newspapers. So while this journalist didn't lie, she left out critical facts and spun those she did report to conform to the story she'd wanted to tell from the outset. For example, since Perelman had been cast as the story's antagonist, he couldn't just be some ordinary home owner who wanted peace and quiet; he had to have some ulterior motive such as being prejudiced against . . . Belgians, because, of course, everybody has it in for those Belgians.[1]

A late-nineteenth-century Tammany Hall politician named "Big Tim" Sullivan once said, "I don't care what the newspapers say about me as long as they spell my name right." Maybe I should be like Big Tim, but I'm not. I try to be ethical and it pains me when people think I'm not. As a council member, I'd worked hard to be squeaky-clean, even turning down a stipend to which I was entitled for chairing the Education Committee, which God knows I could have used since my husband, three children, and I were living in a one-bedroom apartment.

Besides, my reputation was one of the few things I had going for me. I was quite unpopular with the city's unions and the Democratic political machine because I wouldn't toe the party line. For example, I was the sole council member to vote against a law requiring the buyer of an office building to continue employing the prior owner's maintenance workers. I had nothing against maintenance workers, but I didn't see why we should have special employment laws for commercial buildings that didn't apply to other businesses like movie theaters, supermarkets, and apartment buildings. Doing this was a recipe for creating a crazy patchwork of laws.

I was also one of only three council members to vote against a law introduced by my colleague Bill de Blasio, a bright and ambitious council member whom I'd gotten to know when he ran Hillary Clinton's first Senate campaign. His proposed law would allow unions to circumvent the limits on campaign contributions by making them through multiple affiliated entities (such as a local union and its parent). For de Blasio, increasing the power of unions, which would advance his progressive political agenda, trumped the goal of limiting the influence of money in politics. De Blasio's bill would also advance his own career since he'd been fined for receiving more than $20,000 in contributions from the New York Hotel and Motel Trades Council and its Local 6, both of which listed the exact same person as having the authority to decide who received their contributions.

In addition to voting against their bills, I also endeared myself to my colleagues by introducing some of my own. For example, I proposed a law to ban politicians from putting their names on big signs at capital construction projects, since I didn't think politicians should get free advertising on projects for which taxpayers were paying. My bill was about as popular with my colleagues as a ham sandwich at a bar mitzvah.

Fortunately, I was more popular with my constituents since I was attentive to their quality of life concerns. For example, they hated the plastic newspaper racks that littered the streets, so I asked Karim Rashid, a top designer whose clients included Prada and Giorgio Armani, to design new ones. He came up with a version made of silver fiberglass that stood atop mushroomlike bulbous pedestals so it "looked like a growth from the ground" and tilted gracefully backward so that the papers were easier to remove. "Yet another pair of words you're likely to find only in New York: stylish newsracks," commented one paper.

Another big issue was pets. In the summer of 2004, I learned that an angry crowd of dog owners was demanding that something be done about a pit bull that had attacked an eight-pound Chihuahua named Frank. A legal loophole prevented the police from doing anything about dog-on-dog violence, so I proposed legislation to fix it.

But the part of my job that I cared about most was chairing the council's Education Committee. I'd previously been a professor of history and had gotten into politics primarily to improve public education. I'd held dozens of hearings about problems with the public school system. Many newspapers had editorialized in favor of the reforms I'd advocated but the *Times* hadn't, so I was elated when I finally got a call from a member of its editorial board. I returned the call immediately, prepared to wax eloquently about everything I'd learned from my many hearings and school visits. Instead, I was asked about my dog legislation. Oh well.

Of the educational issues on which I'd been prepared to wax eloquently, the most important was overhauling the work rules and job protections for school employees. As schools chancellor Joel Klein observed, "lockstep pay, seniority and life tenure . . . act as handcuffs and prevent us from making the changes that will encourage and support excellence in our system."[2] Klein had proposed removing these handcuffs by adopting an eight-page "thin contract" for teachers. Conditions for reform seemed ideal since the teachers' contract had just expired, I'd just held hearings on reforming them, and we had a mayor, Michael Bloomberg, whose wealth had enabled him to get elected without union support. As one columnist aptly observed, "If not now, when? If not Bloomberg, who?"[3]

On October 20, however, the *New York Post* reported that the city was close to reaching a deal that would allow the United Federation of Teachers (UFT) "to keep the vast majority of the privileges enshrined in past contracts." I was deeply troubled by this as I believed that reforming this contract was critical to improving the school system. I publicly called on Bloomberg to "stand strong." Bloomberg, however, accused me of "grandstanding" and the city council's speaker called my letter "destructive to the process." Ouch. "Eva found out last week what it is like to play in the big leagues," observed a political consultant, "the mayor and the speaker are jockeying for position on education, and Eva got in the way." This wasn't the first time I'd annoyed Bloomberg. He'd called hearings I'd held on shortages of basic supplies like toilet paper "a tempest in a toilet bowl."

Notwithstanding these rebuffs, I felt I'd accomplished a lot on the council. Not only had I passed more laws than any other sitting council member, but I'd also held eighty-eight Education Committee hearings that delved into important issues. Those on the union contracts had drawn national attention and had advanced public understanding of the impact these contracts had on teaching and learning. But those hearings had served an additional purpose. While I appreciated that teachers needed to unionize to level the playing field when negotiating with their monopoly employer, the government, I felt that the UFT had become too powerful, that its ability to give marching orders to nearly every elected official in the city was undermining the quality of education the city's children were receiving. I hoped that by challenging the UFT's hegemony, I could embolden other elected officials to undertake some of the critical reforms that were necessary to fix the public school system.

But to really make the point that opposing the UFT wasn't political suicide, I needed a second act: getting elected to higher office. I'd set my sights on Manhattan borough president and I liked my chances. I'd already gotten elected without the UFT's endorsement—the only Democratic official in all of New York City to do so—and I had a long list of supporters, a willingness to campaign hard, and an appealing résumé as an educator and good-government advocate. Precisely because I had everything going for me except the UFT's support, the race was the perfect test case for whether it was possible to stand up to the teachers' union and live to tell the tale. I'd soon find out.

2

I AIN'T GONNA GET EVA'D

2005

On February 14, 2005, standing on the steps of city hall surrounded by more than two hundred supporters, I announced my campaign for borough president. I'd arrived at the council in 1999, I observed, "with a baby in one arm and a copy of the city charter in the other" and "six years, eighty Education Committee hearings, ten laws, and two more kids later, I am ready to do more."

While I had the advantage of being the most well-known of the nine candidates in the race, I was a persona non grata with the Democratic machine. It hadn't always been this way. During most of my time on the council, I'd had fairly collegial relationships with my colleagues. All that had changed, however, when I'd held hearings on the labor contracts and become the UFT's public enemy number one.

These tensions bubbled to the surface in a *Times* article with the headline "Mayoral Ambitions and Sharp Elbows; Councilwoman Spars Way into a Position of Influence":

> [E]ven as her aggressive, confrontational style has set her apart on a legislative body known less for fomenting change than for renaming streets, it has also alienated many of her colleagues. [A] common refrain is that the councilwoman's ambitions exceed her political skills [and she] fails to . . . build coalitions with other council members.

In fact, my ability to build coalitions was precisely what had allowed me to pass so many laws. My views on the labor contracts, however, were diametrically opposed to those of my colleagues so I'd inevitably alienated them when I'd taken on this issue.

Assemblyman Scott Stringer soon emerged as my leading opponent, racking up the endorsements of many elected officials and of several unions, including the UFT. While none of this was surprising, it meant that I'd need to sweep the newspaper endorsements to win the election. The *Times* endorsement alone was worth about 10 percent of the vote.

I'd also need to campaign hard, so I began a punishing schedule of street campaigning, fund-raising, and house parties. In the middle of all this, my husband, Eric, suggested I visit a school in Queens that a client of his was funding. Eric was a business lawyer who also did a lot of work, mainly pro bono, for charter schools, which are public schools that are run by an independent board. Eric had been a founding board member of one of the city's first charter schools and then had gone on to found another charter school by the name of Girls Prep with a friend of his, Bryan Lawrence. Then, to help spur the creation of more charter schools, Bryan and Eric had held an event for aspiring charter school founders to meet potential donors. They'd gotten help from Whitney Tilson, a hedge fund manager and education reform blogger with a severe case of graphomania and an email distribution list the size of the phone book, and Boykin Curry, a friend of Bryan's who was rich, brainy, ceaselessly gregarious, and owned an apartment with a massive terrace overlooking Central Park that was perfect for such events. More than a hundred people showed up, among them an investor named John Petry.

Petry later asked Eric to help him apply for a charter for a school that would use a reading program called Success for All in which students were evaluated every eight weeks and assigned to small groups with other students who were at the same reading level. Petry's partner, Joel Greenblatt, really liked this program because, like the investing system he was famous for creating, it was simple, practical, and effective. Greenblatt had therefore contributed $2 million to expand the use of this program at PS 65 in Queens. Eric had visited the school and been impressed by it, so he suggested I do likewise. I did and was equally impressed.

Petry was also in the middle of founding an organization called Democrats for Education Reform to encourage Democratic politicians to support charter schools and other education reform efforts. His comrades in this endeavor were Tilson, Curry, and Charlie Ledley, an investor who has since been immortalized in the film *The Big Short* for turning $100,000 into $120 million by betting against subprime mortgages. Their first event took place on June 3, at Curry's apartment, which was quickly becoming the unofficial headquarters of New York's education reform movement. So many people came that Curry had to commandeer a bar downstairs for the overflow crowd to hear the featured speaker, a junior senator by the name of Barack Obama.

As the borough president race entered the homestretch, I campaigned harder and harder, from 6 a.m. until 10 p.m. most days, including several events per evening. We also began mailing our campaign literature. I wanted it to convey my relentless devotion to solving constituents' problems, and my advisors came up with the slogan: "Don't get mad, get Eva." But I knew it would all come down to the *Times* endorsement and on August 28, 2005, they announced their pick. I was "smart and driven," they said, but "abrasive." They preferred Stringer, whom they said had "a sterling reputation as a catalyst for reform."

I cried when I read the endorsement. I felt defeated not just personally but in all that I had worked for civically and educationally. I had tried so hard to be a model public servant: to be ethical and independent, to be principled, to vote my conscience, to tackle issues that really mattered. I'd known this would put me at odds with the political machine and the unions, but I'd hoped I'd at least have the papers in my corner. Instead, the *Times* had gone with the machine candidate.

I refused to give up. I doubled down, as did all of the people who believed in what our campaign was about: Eric, my parents, and our campaign's band of idealistic staffers and volunteers. We worked like maniacs, making tens of thousands of phone calls, campaigning on the street, leaving no stone unturned.

A few days later, just as we were feeling that maybe we could win after all, a mailing went out to voters that pictured a mother with her child saying: "It's hard enough for families like us to get by. And Eva Moskowitz is just making it harder." More negative mailings followed, including one attacking my vote against the building maintenance workers' law. These mailings were sent out by the Working Families Party (WFP), which received $171,000 from the UFT in the two months before election day.

By not counting these expenditures, Stringer was circumventing the limitation on campaign expenditures to which we were both subject. They were also improper because one political party, the WFP, was trying to influence the primary of another party. The Democratic Party's lawyer asked, "What are these guys doing getting involved in our primary race?" He added, however, "It's not this particular race we're interested in."[4] Translation: we hate Eva Moskowitz just as much as the WFP, but it's the principle of the thing!

It was upsetting to see myself being relentlessly vilified, but what bothered me even more was that the *Times* had enabled Stringer to boast of being a "catalyst for reform" while he was doing an end run around the campaign finance laws. I asked the *Times* to withdraw its endorsement but they refused, so I was now up against the Democratic machine, the unions, the WFP, and the *Times*. We fought on, but I knew in my heart our efforts were doomed. On Election Day, Stringer bested us by 9 percentage points.

Bizarrely, a few weeks later, a *Times* editorial called Stringer's abuse of the campaign finance program "pure hypocrisy" and said that if he "wants to make a name for himself as a reformer, he should stand up to fake parties like the one accused of helping him unfairly in his own party primary." But when voters had gone to the polls, the *Times* had told them Stringer had a "sterling reputation" for "reform." Talk about closing the barn door after the horse has bolted.

During the campaign, the UFT had kept quiet about its efforts to defeat me but UFT president Randi Weingarten now publicly boasted that she'd done "everything in [her] personal power, fought day and night" to help Stringer win.[5] Her boasts were effective—from that point on, according to Chancellor Klein, elected officials whom he asked to support his reform efforts would respond, "I agree with you, but I ain't gonna get Eva'd."[6]

Moreover, the next chair of the Education Committee would undoubtedly be in league with the UFT. With me gone, one *Post* columnist asked, "who will be left to ask the questions she's been asking—to spot the accountability-dodging and blame-shifting and simple incompetence that defined the fights over education policy in New York City?"

A month after my defeat, the UFT signed a new contract with the city. While it contained a few modest reforms, such as expediting the firing of teachers who'd engaged in sexual abuse, it preserved the "handcuffs" of "lockstep pay, seniority, and life tenure" that both Klein and I believed were profoundly harmful to the public school system.

For a while, I'd felt like I was having a real impact on politics and public education in New York City, but now it had all come to a crashing halt: I was out, the UFT candidate was in, and the UFT had gotten a new contract with virtually none of the reforms for which I'd advocated. It felt like all of my hard work had been for nothing.

There was, however, one last important service I could render to the city before leaving office: endorsing Bloomberg for reelection. While I was disappointed that he hadn't taken a firm line on the UFT contract, I nonetheless believed he was by far the best candidate. We arranged a joint press conference at which I praised Bloomberg for putting "educational reform front and center" and he said he was "thrilled to have [my] endorsement." When it ended, he kissed me on the cheek. It was strange given all the barbs he'd sent my way over the years, but we had developed a mutual respect. We both cared deeply about public policy, particularly education, and gave as good as we got when fighting for what we believed in. In covering my endorsement, the *Times* noted that I was "widely regarded as the Council's most outspoken advocate for education reform" and that "when a reporter pointed out . . . that [I] had engaged in 'spats' with the mayor over school policies, [I] responded that 'spat might not be strong enough' a term." I hoped this meant the endorsement would carry weight.

Bloomberg won the election and I took comfort in knowing that, as I exited stage left, he'd continue to fight the good fight. I began thinking about what I should do next. While I hoped that I might serve the city as an elected official again someday, I felt it was important to get out of politics for a while, to do something else.

But what?

3

CHAIM'S CHILDREN

1896–1957

My grandfather Chaim Fiderer-Margolis was born in a tiny Polish village named Tluste, where he lived till his family sent him to Vienna at age thirteen in search of used clothing they might resell. He stayed, became a sweater maker, and married a woman who bore him a daughter named Sonia. When that marriage ended in divorce, my grandfather married Sascha Just who, on October 14, 1937, bore him another daughter, my mother, Anita.

During the following years, Germany annexed Austria. "The Hitler Youth spared no effort in molesting me," my grandfather later wrote, "[so] I slept in a different place each night [and] during the daylight hours rode on the trolley cars." Chaim made a vow that if he survived, he would "tell the world of the pain suffered by the Jewish victims," by publishing an account he was keeping in the form of Yiddish poetry. He wrote:

We don't have the visas that the law decrees.
Our road forward is blocked, in vain are our pleas.
So, forsaken and woeful, we languish in prison.
The sentence is harsh; Nazi persecution relentless.

At this time, the future war criminal Adolf Eichmann opened the Central Office for Jewish Emigration, which a contemporaneous observer described as being like "an automatic factory": "you put in a Jew who still has some property, a factory, or a shop, or a bank account" who goes "from counter to counter" and "comes out at the other end without any money, without any rights, with only a passport on which it says: 'You must leave the country within a fortnight. Otherwise you will go to a concentration camp.'"[7] But ironically, however rapacious Eichmann's intent, his Nazi efficiency enabled my grandfather to finally obtain a travel permit which my mother's family used to emigrate to Switzerland in the summer of 1938. There, they stayed in a refugee camp for several years while seeking permission to immigrate to America.

In 1941, they finally got such permission and, on August 6, they boarded the SS *Navemar*, a boat chartered by the American Jewish Joint Distribution Committee. It had begun its voyage in Seville ten days earlier and by the time it reached Lisbon, where my mother's family boarded, thirty-six passengers had contracted typhoid fever and two had died of it. The ventilation was so poor on the *Navemar*, a former cargo ship that had been outfitted with bunks, that most of the passengers had taken to sleeping on deck in the open air.

As the *Navemar* crossed the wide expanse of the Atlantic, conditions worsened. Fresh water was in such short supply that it was rationed and passengers had to bathe with ocean water. Fearing that the food was contaminated, my grandmother didn't let my mother eat much, which proved prescient as it was later discovered that one of the ship's food servers was a typhus carrier. My grandfather wrote:

Filth in every corner,
germs thrive everywhere.
Seven die, are thrown overboard
into the briny deep.

The Navemar sails on
with its cargo of evicted Jews
who hope to reach America
to have freedom and peace at last.

On September 12, the *Navemar* finally reached New York Harbor where it was inspected by a team of doctors and sanitary inspectors led by Medical Director H. F. White, who wrote that he'd "never observed a vessel arrive under conditions so insanitary and so fraught with potential danger." Had it been necessary to batten down the boat's hatches due to bad weather, he observed, many passengers would have suffocated due to inadequate ventilation; had the boat sunk, most of its lifeboats could not have been launched.

But fortunately, neither event had transpired and when my mother's family stepped off the *Navemar,* they were confronted with a skyline like none other. New York City could boast of not only the world's tallest building, the Empire State Building, but also an astonishing thirty-one skyscrapers that had been completed that same year. Although my mother was just three years old, she vividly recalls standing with her family at the bustling wharf, surrounded by all of their worldly belongings, as her father cried out the last name of the relative who was supposed to meet them: "Feederer . . . ! Feeee-der-er!" Eventually, his sister Minna appeared and took them to the Hebrew Immigrant Aid Society.

While New York was a foreign land, it was also a city of immigrants just like my mother's family: Irish who'd fled the great potato famine, Italians who'd fled poverty, Armenians who'd fled genocide, and Jews who'd fled pogroms and now Nazis. Two million of the city's residents were of Jewish heritage including the city's mayor, Fiorello La Guardia, who, like my grandparents, spoke German and Yiddish.

After living briefly on Manhattan's Lower East Side, my mother's family moved to Washington Heights where my mother remembers playing in the playground and riding on a pony in Fort Tryon Park. Around this time, in July 1943, my grandmother went to a doctor because she'd noticed she was losing weight. My grandfather recorded what followed:

> On this beautiful summer's day, Sascha came back home, sent Anita out to play, and with a restrained but tearful voice said: "Mo, I have a tumor in my stomach."

On August 29, I told Anita she has to go to a home since Mama has to go to a hospital. She asked, "How long will I stay there?" I said "Only for two weeks." She took this at face value. "Alright, I will go to the children's home so I will have a healthy mommy."

[After an operation] Dr. Steinhardt told me: "Herr Fiderer. We took out two tumors and both ovaries. We could not remove anything on the bladder because without a bladder one cannot live. We will try to treat her with radiology." I asked: "Frau Doctor. Does my wife have cancer?" She answered, "Yes, Herr Fiderer. We have done everything possible. You have to pray to God. God can still help."

Sascha opened her eyes and asked, "what did they do to me?" I told her that both tumors had been removed and now she will get better. She took my hand and kissed it.

On this tragic evening, I went home full of despair. I was not prepared for this blow of destiny.

September 5. For the first time since Anita was in the home, I visited her. She asked, "Papa, how long is two weeks? You told me I will go home in two weeks." She sent for Mommy a thousand kisses.

December 5, 1943. Dr. Steinhardt reported "Unfortunately, the radiology was not effective."

January 25, 1944. When I visited Sascha today, she surprised me with a fountain pen and sang Happy Birthday to me. The urge to live is so great. All the doctors gave up on her, but Sascha hasn't. She wants to live in spite of weighing 76 lbs. down from 137 lbs.

February 8. [Sascha] urged me to take her home as she couldn't stand the hospital any longer. That evening I packed Sascha's clothes. Two Saschas could fit in these clothes, I thought.

March 26. Eight months ago, how blooming she looked, full of hope, animated. The newly furnished home, every corner clean, the child nicely dressed, and she only thought for the well-being of the family. The dear child with strange people. When I call her,

her first question is, "How is Mommy? Every day I ask God that Mommy shall be healthy." Oh Destiny, how merciless you are.

A month later, Sascha died. My grandfather wrote:

The earth covers only the body;
It can never cover what is deep in one's heart.
The well of tears will someday dry up
But nothing can heal the pain in my soul.

Since it was unthinkable in those days for a man to raise a daughter alone, my grandfather took a third wife, Henriette Strassman. My grandfather had quit a job he'd held at a clothes factory and gone into business for himself, making sweaters that he'd sell at the Essex Street Market, an indoor market created by Mayor La Guardia to replace the pushcarts that had previously clogged the streets of Manhattan's Lower East Side. My mother contributed by manning my grandfather's stall on weekends and sewing buttons on sweaters, earning a penny a button. As my grandfather's earnings were meager, Henriette took a job as a bookkeeper at a local Singer sewing machine store.

My mother, nicknamed Rusty because of her red hair, attended both public school and Hebrew school and spent her summers at Ein Harod, a Zionist camp that prepared children to immigrate to Palestine by teaching them how to set up tents and make wood fires. She went on to attend Music and Art High School and then studied art at Cooper Union, a free college from which she graduated in 1957.

After graduating, my mother worked as a waitress at a resort in the Catskills, an area in upstate New York that was so popular with Jews it was dubbed the "borscht belt" after the beet soup that is popular with European Jews. Uncertain what to do with her life and hearing that a classmate had applied to the University of California, Berkeley, my mother decided to do the same and was accepted.

4

THE ULTIMATE CHARTER SCHOOL BAKE-OFF

2005–2006

Shortly after losing the borough president race, I received two job offers: running the charter school that Joel Greenblatt and John Petry were starting and leading a DOE teacher training program for which Chancellor Klein had raised $20 million. While Klein's offer was appealing because it would have a broad impact on the public school system, it would be run out of the City University of New York, and it soon became clear they had me pegged for the role of figurehead in chief. They wanted to bolster their budget by using Klein's money to pay the staff they already employed to teach the courses they already taught.

If I ran a charter school, I'd have real control that would enable me to strive for true excellence. I also liked the idea of working with Joel and John; they were razor-sharp and really wanted to help kids. Moreover, their longer-term vision was bold: figuring out how to run a school that cost no more than the district schools but got far better results, and then replicating that model over and over dozens of times. Cracking that nut could revolutionize American education.

I also liked the idea of getting back into education. I'd always loved the thrill of exposing students to new ideas, and visiting hundreds of schools as chair of the Education Committee had sparked my interest in returning to the classroom. I'd spent years thinking, writing, and talking about K–12 education; now I'd have a chance to apply what I'd learned.

I'd also become increasingly comfortable with the charter school concept. Charter schools were just another kind of public school: they were paid for with government funds, were approved and overseen by government entities, and in some respects were actually more egalitarian than district schools, which had zoning and admissions policies that made them highly stratified by race and socioeconomic status. By contrast, charter schools admitted students by random lottery.

So, I accepted Joel and John's offer to lead what would be called the Harlem Success Charter School and they asked me to draft a proposed contract. I realized this was an opportunity to determine the contractual terms under which the school would hire its first employee—me. While executive contracts often contain provisions for severance or guaranteed periods of employment, I felt strongly that the school's employees should be fully accountable for their performance and that this should start with me. Thus, I turned down Joel and John's suggestion that I draft a proposed contract and informed them that I would instead "serve at the pleasure of the board as an at-will employee."

Now I had to figure out how to open a school in just eight months even though we had no staff, no teachers, no principal, no facility, and only a partially fleshed-out curriculum. Moreover, I was going to be doing this under a very bright spotlight as the *Times* soon observed:

> Opening a school is a risky thing for an aspiring politician; in doing so, Ms. Moskowitz is laying her mayoral ambitions at the tiny feet of 5-year-olds. Even the best-run schools are full of potential surprises—the troublemaking student, the errant teacher, the irate parent—and Ms. Moskowitz, a mother of three who has visited hundreds of schools throughout the five boroughs, knows as well as anyone how hard it is to get it right.

George Arzt, a prominent political consultant, commented, "Your political career is in the hands of the students and the faculty, and that would make me an insomniac."[8] Randi Weingarten, who was still gloating about ending my political career, was apparently looking forward to another helping of schadenfreude. If my charter school failed, she asked, "Will she do what she does now, which is blame others, vilify everybody else? Or will she take responsibility for what goes wrong?"[9] In her eyes, I'd decided to build a house of glass after four years of throwing stones.

To raise the stakes even higher, the UFT had just opened its own charter school, which Weingarten had promised would "dispel the misguided and simplistic notion that the union contract is an impediment to success" with "real, quantifiable student achievement." The UFT and its archenemy were opening schools virtually simultaneously with starkly different philosophies; it was the ultimate charter school bake-off.

I set about assembling a team by asking the battle-hardened idealists who had worked with me on my campaign and on the council to join me in this new adventure. Most agreed, including Jenny Sedlis, a veteran of both my district office and several of my campaigns, Sheila Lopez, my Education Committee policy analyst, and Thomas Melvin and Sarah Szurpicki, who'd both worked on my borough president campaign. In addition, Joel Greenblatt had recruited PS 65's principal, Iris Nelson, as an advisor.

There was one person, however, whom I had to let go:

> Dear Eric:
>
> As you know, I recently assumed responsibility as Executive Director of Harlem Success. I know I speak for the Board when I say I am enormously grateful for your advice and assistance. Unfortunately, I must terminate our business relationship.
>
> Love,
> Your Wife
> *Eva Moskowitz*

Eric had actually told me I had to fire him because, even though he'd been hired by Joel and John, he knew I'd be accused of nepotism if I kept him on. So I took Eric's advice, but I also wrote:

It is my hope that you will consider being our pro bono legal counsel. Though not financially rewarding, we are a hell of a lot of fun.

Happily, he agreed.

While the plan was for our schools to eventually break even on public funding, we needed money for start-up costs such as furniture, books, and our salaries prior to the school opening. Moreover, our school would initially run a deficit since our funding would be based on enrollment, but some of our costs, such as the salaries of the principal and the office manager, would be fixed. I remembered that when I'd held my labor contract hearings, I'd gotten a congratulatory call from the founder of the Gap, Don Fisher, so I arranged to meet with him. He warned me in advance that he ordinarily only gave money to charters with proven results, but he warmed up to the project when we met. He liked that we were starting with kindergarten because, after years of funding charter schools that started in middle school, he'd concluded that it made more sense to begin earlier. In the end, he agreed to make a tremendously generous $1 million contribution to help us get started. This, in addition to the $1 million that John and Joel were giving annually, put us on sound financial footing.

Now I turned to planning the school. Many charter school founders focus on closing the racial achievement gap but I was also worried about another gap: that between American students and those in countries such as Japan, Singapore, and Finland. Closing that gap would require a very high level of rigor, particularly in math and science. In addition, while many of our students would come from poor families, I didn't want to design a school that served one class of students, but rather one to which any parent would be proud to send their children.

Being older than most charter school founders, I already had three children, so I thought about what type of education I'd want for them, which made my approach less ideological. Many educators debate what's most important—conceptual understanding or precision, the rules of grammar or the ability to express one's own views—but kids need all of these things.

The Success for All reading curriculum was the cornerstone of our program. To fill out the rest of it, I drew from my work on the Education Committee. I'd noticed that science wasn't taught until middle school in the district schools and, even then, they failed to teach the scientific method of formulating hypotheses and testing them. This was a lost opportunity. Not only do young children develop their critical thinking by studying science, they also love hands-on activities. I therefore decided to have dedicated science teachers provide instruction five days a week, beginning in kindergarten. Finding a curriculum was the next challenge. Many science textbooks were rudimentary and had boring activities such as having students sort fabrics by color and texture. We hired a teacher from Brearley, a renowned private school, to help develop our science curriculum. She concocted more interesting experiments, like determining which fabrics were hardest to clean after being stained with ketchup and mustard!

Remembering how much my own mother had read to me and believing it had played a critical role in my education, I decided to require that parents read to their children and keep a log of what they'd read.

Like most charter schools, we planned to have a longer school day, but rather than devote all the extra time to academics, we intended to use it for sports, music, art, and especially chess—a game Eric had taught our eldest son to play in order to develop his ability to concentrate and think strategically, and which I felt had also boosted his self-confidence. In addition, we decided to have a school uniform to send the message that we expected our students to take school seriously and also so they wouldn't have to compete to wear the coolest clothes.

Perhaps most important, we needed a school building. Thankfully, New York City had a policy called "co-location" that allowed charter schools to use space in underutilized district school buildings. The city agreed to put us in PS 154, a school operating at 50 percent capacity, but when the UFT got wind of this, they began whipping up opposition. Jenny Sedlis and an assistant attended a meeting regarding our co-location. An aide to a local politician claimed that co-locating us would cause class size at PS 154 to triple. Racial epithets were hurled at Jenny; her assistant, who was African American, was told "to go back to the suburbs." A motion was then made to eject them, which was accompanied by shouts of "throw 'em out."

Two weeks later, the UFT followed up with a rally outside PS 154. The UFT had just succeeded in ending my last career. Maybe they'd derail this one too.

Then I suddenly found myself at war on a second front. I'd decided to add the word "Academy" to our name to convey that our school would be academically rigorous. Technically, this required approval from the State Education Department (SED), but since they were responsible for regulating thousands of schools, many of which had very serious academic problems, they hardly had time to worry about something like this. Or so I thought. SED claimed that since one of our board members had observed that we might sometimes call ourselves Harlem Success Academy for short, thus leaving out the words "charter school," our name change was a nefarious plot to hide the fact that we were a charter school. I protested to SED:

> Most schools have a nickname. Girls Preparatory Charter School calls itself Girls Prep. My alma mater, the University of Pennsylvania, calls itself Penn. Harvard University frequently refers to itself simply as Harvard. [N]o one has to my knowledge accused them of trying to hide the fact that they were universities.

SED eventually caved and agreed to our name change. Now maybe you're saying, "Okay, Eva, they were wrong, but why did you get your knickers in a twist? After all, a name change is pretty trivial." That, however, was precisely the point: if I'd allowed SED to get involved in something as trivial as a name change, who knew what they'd try to meddle with next.

In Thomas Hardy's novel *Far from the Madding Crowd,* a shepherd has a young dog who one night chases the sheep so zealously he runs them right off a cliff. The dog's error, wrote Hardy, was thinking that "since he was kept for running after sheep, the more he ran after them the better." Unfortunately, SED's regulators had fallen prey to the same error of logic. They believed that the more heavily they regulated the better they were doing their job. In fact, regulation is necessary but should be undertaken judiciously, particularly in the case of charter schools since they are supposed to be freed from micromanagement. SED, however, apparently hadn't gotten the memo.

Meanwhile, I sought to recruit students for our school by visiting day care centers. At one, a woman named Natasha Shannon said to me: "So let me get this straight. You have no building, no principal, no teachers, but you want me to enroll my child in your school?" But despite her understandable skepticism, she did enroll her child, who is now a junior at our high school.

Another challenge was recruiting teachers. Since our founding faculty would set the tone for the school, they had to be strong, so we did everything we could to find them: advertising, going to job fairs, reaching out to our personal networks. Once we found candidates we liked, I tried to make up for the fact that we didn't have a school by inspiring them with the excitement of starting something new and with our bold vision. It wasn't easy, but we did manage to hire twelve strong teachers, and while not having a school made hiring harder, those who did sign on were unusually adventurous and idealistic. We also recruited Paul Fucaloro, a veteran teacher at PS 65, the district school Joel had funded.

Meanwhile, I tried to overcome the opposition to our co-location. One local politician suggested it would help if I hired a well-connected insider as an assistant principal. I politely declined. I wasn't going to turn Success into a patronage mill. A few weeks later, DOE deputy chancellor Garth Harries asked us to accept fewer rooms at PS 154 than we'd previously been offered, but I couldn't agree to that. The original offer had split the space fairly between Success and PS 154 based on enrollment; I wasn't going to let the UFT bully us into taking less. Harries pushed back. He claimed that if my child was in a public school where DOE was trying to co-locate a charter school, I'd be opposing it too. I replied that it just so happened that my oldest child was in such a school and I'd refused the entreaties of the other parents to oppose the co-location.

DOE then offered us an alternative site: PS 149 at 118th Street and Lenox Avenue. Fortunately, the UFT chose not to raise a ruckus at this placement, perhaps realizing that if they prevented me from opening a school, they'd deprive themselves of the satisfaction of seeing me fail. Conveniently, Eric and I had just moved to a building in Harlem that was just two blocks south of PS 149.

We had decided to renovate the space we were given because a school's appearance sets the tone for a school, particularly a new one. It was a frantic rush to get this done by August 21, the day we were set to open, so contractors were practically tripping over one another. Then one day we found men with guns hanging out in our hallway. It turned out they were plainclothes police officers who were conducting surveillance on a drug kingpin across the street. Since the presence of armed men didn't really mesh with the safe atmosphere we were seeking to convey to our families, we asked the officers to conceal their weapons.

On August 18, the mother of one of our children informed us her son wouldn't be coming. She lived in the Bronx and her sister, whom she'd been counting on to take her son to school each day, had been murdered. I'd figured our school would be touched by violence, but not before it even opened.

As the first day of school approached, I became increasingly conscious of the difference between debating educational policy in the abstract and running an actual school. It reminded me of the sense of responsibility I'd felt when I'd given birth to my first child and realized that for the next eighteen years I'd be waking up knowing that he was my responsibility. Now I'd be responsible for 165 children.

But while I was nervous, I was also excited and optimistic. We'd planned carefully, created a great curriculum, and hired wonderful teachers. We had a shot, I believed, at accomplishing something truly extraordinary. Filled with this heady froth of hope and terror, I rose early on the morning of August 21, walked the two blocks from my home to our school, and waited for the students to arrive.

5

WEEVILS!

2006

At 7:15 a.m. sharp, we opened the doors and in streamed 165 children dressed in our orange and blue uniforms looking happy, energetic, and full of potential. I felt a sudden urge to confess to their parents that while I was going to try my best, I'd never run a school before and didn't really know what I was doing. Instead, I put on a confident friendly smile, introduced myself, and shook the hand of each child who entered.

Our first task was to match the students with their teachers and, for all of our planning, we'd given this little thought and chaos ensued. Hours later, we discovered we'd mixed up two identical twins.

The days that followed were like some *Alice in Wonderland* dream. Periodically, loud bells would ring throughout the school, their origin and purpose a mystery. We soon learned it was PS 149's system for calling the custodian, as if cell phones, beepers, and walkie-talkies hadn't been invented. In addition, our bathrooms lacked toilet paper, most of our families didn't turn in logs of the books they were supposed to have read to their children over the summer, and our payments from the city were late, so we had to get an emergency loan from Joel so our teachers' paychecks didn't bounce.

On the second day, Paul and I had to serve lunch because most of the cafeteria workers didn't show; we had to cancel recess because the play yard was strewn with glass from broken bottles; the school nurse called in sick; our office's vintage air conditioner broke down when temperatures were in the nineties; the mother of one of our students was checked into a psychiatric hospital; and while the toilets now had toilet paper, most had gotten clogged up from actual use.

On the third day, families told us the nearby public libraries were refusing to lend them books because they were borrowing too many. Librarians against excessive reading—who knew?

Wonderful things were happening too, but the pace of events was overwhelming. Here was a report I gave to Joel and John about the next two days:

Good news:

Chess started on Monday. The kids love Mr. Sanchez.

Kids love science!!! They ate herbs today. Only problem was that they reported to parents that they were eating leaves at school and we had to field some concerned parent calls.

Our kids have read 4,626 books!!!!!! 79 percent of our families completed the reading assignments up from 37 percent.

We call when children are not in school by 9 a.m. and we have reduced latenesses from 20 to 2.

Bad news:

We have 2 weak 1st grade teachers. One is extremely emotional (cries every day) and they are inefficient instructors. They don't know how to use every moment.

[A student] threatened our learning specialist today. [He} made motions to hit her and then jumped from a desk and pulled her hair.

We had another teacher cry today. She lost it. The kids were disobeying. She has 7 special ed boys who misbehave on a regular basis. I am finding a way for more recess. Boys have trouble w academic culture: too much sitting still.

Milk was frozen. Kindergartners cried. Mayhem resulted. Spoke to Miss John. Milk has to be liquid!!!!!

Electricity went down again. Wireless doesn't work all the time. Rugs were dirty. Had to have a heart to heart with custodian which was: if my rugs aren't vacuumed, you're a dead man.

Soccer games: Reserving fields is challenging but we have 3 games scheduled.

Got to bottom of mystery [of] harassing phone calls: a parent. Brought in police. Good God.

And the hits kept on coming. When water started pouring into our office out of an adjacent bathroom, we had to disconnect our newly purchased computer equipment and move it to higher ground. After flooding came pestilence: the snacks DOE had given us for our kids were infested with weevils.

Many of our students still hadn't received their uniforms, so we fired our uniform provider. Eleven families told us they couldn't read even a simple picture book to their children because they were illiterate or didn't speak English. Other parents were belligerent, including a woman who used a string of profanities to tell me what a terrible school I was running since we'd let her grandson lose his gym uniform, a reprimand I took personally since things were going so poorly.

By the end of the second week, I was exhausted. We'd had a parent with a nervous breakdown, broken glass on our play yard, parents not reading to their children, an incompetent uniform company, failing electricity and Internet, a librarian work slowdown, a broken air conditioner, belligerent parents, nonworking toilets, a police stakeout, a cash crisis, a sick nurse, frozen milk, and weevils. Weevils!!!

If this was just two weeks, what would a year be like? And what would it be like when we had not 165 students, but 500 or, God forbid, dozens of schools with thousands of kids as Joel and John wanted? Maybe I wasn't cut out for this work. Maybe I'd been naïve to think I could run a school with so little experience.

But I couldn't just fold up my tent, so we went about solving our problems one by one. To get rid of the bell system, we bought walkie-talkies for the custodian and both schools. To end the New York Public Library's policy of discouraging excessive reading in Harlem, I called its president, Paul LeClerc, who sent down word that this wasn't in fact official library policy. To help a family whose parents weren't reading to their child because they'd recently become homeless, I called up the commissioner of housing and got them a housing voucher. To address our nurse's habitual lateness and inability to identify common childhood ailments like ringworm, I got a new nurse assigned. These successes boosted my spirits. While I was inexperienced, at least I could use my knowledge of city government and the problem-solving skills I'd learned as a council member to support our teachers and help our families.

Next, I spoke to the grandmother of the boy who'd lost his uniform. Her anger, I came to understand, stemmed from anxiety that she'd have to buy a new uniform every week, which she couldn't afford. I said we'd pay for a replacement uniform and do a better job of helping her grandson keep track of it in the future. Given how she'd treated me, however, I feared she might act belligerently toward our teachers. I explained to her that we really did have her grandson's best interests at heart, that we were all on the same team, and that in the future, she needed to give us the benefit of the doubt. It's very important that a school protect its teachers from being mistreated by parents or they become jaded and dispirited, which ultimately hurts kids.

I also worked hard to improve our school's appearance. A poorly maintained facility suggests that teachers and students aren't valued and contributes to low expectations. If only half of a school's light bulbs work, teachers figure it's okay to prepare for half their classes and parents figure it's okay for their children to do half their homework. I wanted our facility to communicate that our school was both rigorous and joyful, and to instantly dispel any preconceptions anyone had about urban public schools. This meant having classrooms that were spotless and well lit, hallways that were festooned with student work and colorful banners, and bathrooms that both looked and smelled clean.

Success eluded me at first: I'd complain to the custodian; he'd promise improvements; none would come. Finally, I asked him whether there was anything I could do to help. Yes, he said, I could buy him a better vacuum cleaner, which he probably thought was impossible since district school principals didn't have the authority to do something like this. I did. Done, I told him. Second, he wanted us to put students' chairs on the desks at the end of the day to facilitate cleaning. Done, I said. Third, kids had to stop messing up the bathrooms by throwing paper towels all over and failing to "aim." He was 100 percent right: why should his staff put any effort into keeping our school clean if our students' conduct suggested they didn't care if it was clean? We began monitoring our students' conduct in the bathrooms more closely and instituted a bathroom cleanliness competition between the boys and the girls with the winning gender receiving a golden plunger award. The cleaning gradually improved and we helped the custodian hold his staff accountable by regularly reporting on the quality of their work. We also thanked them for their efforts and this helped them understand that their work really mattered to us.

Another problem we had was getting access to our building's library, one of dozens the Robin Hood Foundation had built for about $1 million a pop. These libraries were so wonderful that rather than waste them on students, many schools turned them into de facto teachers' lounges. I got access for our students by threatening to contact Robin Hood, but PS 149's students had no such luck. This illustrates the danger of indiscriminately providing additional resources to schools; while they may benefit a competently managed school, a dysfunctional school will merely squander them.

My highest priority was creating a school culture that had a low tolerance for laziness and dysfunction and high expectations for student achievement and teacher performance. To accomplish this, I drew on lessons from other pioneers in the charter movement including KIPP, Achievement First, and Uncommon. I learned from them that focusing students on the goal of attending college would help them understand that doing well in school wasn't just about pleasing one's teachers but about having a bright future. We took our kids to college campuses starting in kindergarten, referred to grades by the year they would graduate from college (e.g., our kindergartners were the class of 2023), named our classrooms after our teachers' alma maters, and encouraged our teachers to talk about their college experience and to decorate their classrooms with swag from their alma mater.

We also preached what we dubbed our "ACTION" values—Agency, Curiosity, Try and Try, Integrity, Others, and No Shortcuts. These may sound hokey but they really strengthened our school culture. Agency reflects the idea that students need to take responsibility for their own fate rather than simply rely on others; curiosity, our belief that learning should be joyful and driven by natural intellectual drive. To teach these values, we used games, stories, and songs. We taught agency with a song called "I'm in Charge of Me." We taught try and try with a Dr. Seuss book in which a boy creates letters beyond *Z* to capture the idea of going beyond what is ordinarily considered sufficient. While it can be frustrating to teach young children because they don't know how to behave, the upside is that they are virtually a blank slate, and if you take advantage of that fact to teach them to become good learners, that investment will pay dividends for years to come.

Our ACTION values also applied to our adults and, to express our dismay when adults failed to exemplify these values, we came up with their antithesis which we called, tongue in cheek, LAGDEC: Learned helplessness, Apathy, Giving up, Dishonesty, Ego, and Corner cutters.

By a couple of months into the school year, I felt reasonably good about the culture we'd created and our school no longer seemed out of control. But it became increasingly clear that I'd made one critical mistake. The principal I'd hired wasn't stepping up to the plate despite my best efforts to help him, so in October, I let him go. This meant I'd have to hire a principal midyear, and given that I hadn't managed to succeed in finding a strong candidate in the normal hiring cycle when candidates were plentiful, it seemed unlikely I'd do so now when candidates were scarce.

6

THE FUCALORO METHOD

2006–2007

Given the difficulty of hiring a principal midyear, I considered promoting one of my teachers but, as talented as they were at instruction, none were ready to take on the responsibility of leading a school. I had no choice, I concluded, but to do the job myself. I knew I wasn't an ideal candidate since my only experience teaching children was a history class I'd taught one summer to fourth-graders, but I'd learned a great deal about K–12 education by holding innumerable hearings and visiting hundreds of schools as chair of the Education Committee. It was like getting a graduate degree in education.

But given my lack of actual classroom experience, I decided to focus my energies at first on supporting our teachers: providing them with a clean and well-maintained facility, adequate supplies, counseling for students who needed it, and, perhaps most important, helping with the discipline issues that often confounded and demoralized teachers. While a competent teacher can handle most behavioral problems, some students misbehave incessantly and fail to respond to ordinary classroom management techniques, which can have a domino effect; when other students see it's possible to get away with misbehaving, they decide to get in on the fun and soon the teacher is playing whack-a-mole rather than teaching.

A school often needs parental support to improve a child's behavior. If the school tells the child he must behave better but the parents say he needn't, the child will invariably choose the latter course. It's often hard, however, to get parents on board. As a parent myself, I know how easy it is to be defensive when you're told your child is misbehaving. It feels like your child is being attacked so your instinct is to defend him, to say 1) he didn't do it; 2) if he did, the other kid started it; 3) he never misbehaves at home; 4) the teacher is playing favorites; and 5) the teacher doesn't know how to manage her class. And of course, sometimes these things are at least partially true: some teachers do have weak disciplinary skills and some kids do find it harder to behave in school than at home since there are different behavioral expectations in a classroom with thirty students.

There's also an issue of trust. Since district schools are often quick to send children with behavioral issues to programs for emotionally disturbed children, some parents fear that admitting their child has a behavioral problem will merely grease the skids. It's critical to help parents understand that addressing a student's behavioral issues is important not only so that the school can be more orderly, but also so that the student can succeed academically. You must make clear to a parent that you don't see their child as just some two-dimensional "bad kid" but rather you see his potential. I could do this, I discovered, by finding out something positive about a student, such as a talent for science or art, and talking about that first. This is quite effective, but it's hard to do when a child is being challenging. Sometimes you want to grab a parent firmly by the lapels and scream, "Your child is making my job impossible!" But unless you acknowledge a child's virtues, a parent will rarely admit his faults.

Occasionally, parents would object to our imposing our cultural expectations on their children and some of our white teachers were hesitant to push back because they feared doing so would be insensitive. I found, however, that if I told parents we had the same high expectations for all students and I was confident their children could meet them, they were actually pleased. This may be counterintuitive, but ironically, just because a parent felt we might be willing to set a different standard for a student's conduct based on his race or class, that didn't mean they actually wanted us to have that view.

Once we got parents on board, their job was to work with us to address the behavioral issues. We'd give them a daily report on their child's behavior; if it was good, they'd say something positive to their child such as, "I heard you behaved very well in school today. I'm so proud of you." If the report was bad, they might say, "I heard you pushed another student. If you get frustrated, you need to talk to the teacher so she can help you calm down." We weren't looking for parents to take extreme measures; indeed, we sometimes had to discourage them from doling out physical punishments.

Rather than overwhelm parents by trying to solve every problem at once, we aimed for small victories to build momentum. When one student who liked to jump down a flight of stairs in a single bound managed to walk down the steps one by one, I took a picture and emailed it to his parents.

To reward good behavior, we'd call up a child's mother in the middle of the day so he could speak to her by phone. We'd also call families on Sunday nights to tell them what would be happening in class that week, such as what stories would be read, as we found that knowing what to expect had a calming effect on children.

We weren't seeking to foist behavioral problems off on parents. Rather, we were asking them to work with us as a team so we could present a united front. We wanted the child to understand that he would incur the opprobrium of all of the critical authority figures in his life if he misbehaved, and, if he didn't, their praise. This team effort, combined with incentives we provided for good conduct, usually worked quite well.

With these efforts, our classrooms gradually but surely became calmer and more productive, which helped the morale of our teachers. So did being attentive to their needs. When some of our teachers mentioned they were having trouble staying on top of their work, we hired an assistant to correct homework, take attendance, and help out with other administrative tasks. We also made sure our teachers got any supplies they asked for immediately, which surprised them since they'd often had to pay for such supplies out of their own pockets in the district schools where they'd previously taught. Spending a few hundred dollars for supplies that made our $60,000-a-year teachers more effective and enthusiastic was a no-brainer.

I next turned my attention to instruction. Since people viewed me as a politician, many assumed I'd be a figurehead who wouldn't get my hands dirty with actual schooling, but I was first and foremost an educator. To learn more about instruction, I decided to teach an eight-week session of the Success for All (SFA) reading program. I attended the planning sessions led by Paul Fucaloro, whom our teachers tended to underestimate because he bore a superficial resemblance to the movie stereotype of a stern and joyless teacher. In reality, Paul was nothing of the sort. He was an avid gardener, a great cook, and a manager of professional boxers—something I didn't quite believe until Paul had half a dozen of his protégés show up one day to help us move boxes. Paul always had some business scheme going on. Once, he announced he was heading off to Dubai to buy crude oil from a sheikh he'd befriended through boxing; he returned with the news that the crude oil deal had fallen through but he'd been able to buy a vast quantity of olive oil instead. His energy was limitless. I'd show up at 6:30 a.m. to find not only that Paul had already been preparing for an hour but that he'd also planted his annuals that morning before coming in.

As for teaching, Paul could play every position on the field. Since we didn't have a music teacher our first year, he led sing-alongs of popular '60s and '70s tunes like "Under the Boardwalk." When we needed materials to help students with reading comprehension, Paul wrote marvelous poems designed to teach children skills such as identifying a passage's main idea.

I was struck by how thoroughly Paul prepared and how much he enjoyed doing so. Some teachers complained that SFA wasn't sufficiently challenging, but Paul showed me this was because they were focusing only on phonics rather than on comprehension. For example, if a character who hadn't been invited to a birthday party said he didn't want to go anyway, the students would take this at face value, not realizing that authors often expect their readers to deduce a character's true feelings. Paul would ask the class whether the story contained any clues that the boy might feel differently than he claimed.

Of the many things I learned from Paul, the most important concerned effort. Many teachers assume that students are generally trying their best. Alas, that's rarely true. Imagine how carefully you would add up ten numbers if you knew you'd get $1 million for doing so correctly. You'd focus ferociously and check your work ten times to make sure you got it right. Now let's define that as your best effort. By that standard, how often do you truly give your "best effort" when you do things? Most adults rarely give anything even close to their best effort and kids are even worse. Paul knew this and he believed that it was his responsibility to get kids as close as humanly possible to their theoretical million-dollars-on-the-line level of effort every minute he taught.

Here is how this would play out. Suppose the boy in the above-mentioned story had slammed the door on hearing he hadn't been invited to the birthday party. Paul asks, "How do we know that Kevin is angry?" and Elijah answers, "Because Kevin wasn't invited to the birthday party." Many teachers would follow up with "Very good, Elijah, that's *why* Kevin's angry. Now, how do we know he's angry?" This may seem very pedagogically astute because the teacher compliments the student for what he did right while still pointing out the error. However, Paul would start from the premise that the student probably hadn't given his best effort and would press the student to try harder. He might say, "Elijah, you didn't listen carefully to my question. Aisha, repeat my question so Elijah can try again." This may be tough medicine but Elijah will now try harder. Moreover, Paul gives Elijah a chance to redeem himself by trying to answer the question after Aisha restates it rather than just asking Aisha to answer the question.

Children are less fragile than many educators think. They can handle being pushed hard intellectually. Moreover, since children saw that Paul was tough on all of his students, they understood that if he was displeased with them, it didn't mean they were stupid but that Paul had high expectations. In fact, Paul's implicit message to his students was that he thought so highly of their abilities that when they failed to answer his question correctly, the only explanation could be lack of effort. Students invariably rose to the challenge in the end and this led to real self-esteem, which comes from hard work, not false praise. That's why students who play sports are often more confident than their peers: their coaches push them so they become accustomed to the idea that they can surmount challenges with hard work.

Because Paul was demanding, some teachers thought he was mean, but students adored him and worked particularly hard in his class. Kids like predictability, and Paul, while strict, was utterly transparent: kids knew exactly what conduct would garner censure and what conduct would earn them praise.

Students also enjoyed Paul's class because he'd perfected techniques to keep them constantly engaged. When Paul asked a question, he'd pause briefly so that every student could prepare an answer and then call randomly on a student, who was expected to have his answer ready. Thus, every student had to prepare to answer every question. When the student gave his answer, Paul would sometimes ask another student to repeat it to make sure they were listening to one another. Paul never repeated a student's answer, as many teachers do, because that just encourages students to stop listening to one another. Here is a typical sequence of Paul's questions (with the students' responses omitted):

"Was Kevin telling the truth when he said he didn't want to go to the birthday party?"

[*Pause to allow every student to prepare an answer.*]

"Elijah?"

"Elijah, what evidence is there in the story for your opinion?"

"Aisha, repeat what Elijah said."

"Raise your hands if you think that Aisha correctly stated Elijah's opinion. Now, raise your hands if you think she didn't."

"Taj, how was what Aisha said different from what Elijah said?"

"Rashid, do you agree with what Elijah said? What is your opinion? What evidence is there for your opinion?"

"Raise your hand if you agree with Rashid."

"Mark, you didn't raise your hand. Why do you disagree with Rashid? What's your evidence?"

Students knew they could be called upon at any moment to answer Paul's rapid-fire questions and were expected to have their answer ready. As a result, they continually engaged in "active listening": developing their views on everything that was being said. The moment a child made a mistake, a dozen students would start squirming in the hope they'd be called upon for a correction. What Paul realized was that students get nearly as much intellectual benefit from preparing a potential contribution as from actually making it, so children could learn a lot even if they rarely got a chance to participate. This epiphany was the foundation of what I've dubbed the Fucaloro method, Paul's riff on the Socratic method.

Paul made sure students were engaged in active listening by closely monitoring their behavior. He didn't let them stare off into space, play with objects, rest their head on their hands in boredom, or act like what Paul called "sourpusses" by bringing an attitude of negativity or indifference to the classroom. Rather, they had to sit up and "track" (meaning look at) the speaker so Paul could be sure they were paying attention. If a child's eyes strayed for a moment, Paul would point to the child, snapping his fingers if necessary to get the child's attention. Students who were short or were strategically slouching to avoid eye contact were moved to the first row. If

Paul had thirty students in his class, he wanted to see sixty eyeballs staring back at him.

To improve my own teaching, I hung on Paul's every word, followed his advice religiously, and went to him for extra help. At the end of eight weeks, my students were assessed and they'd made excellent progress. This proved to me that even an inexperienced teacher like me could be fairly effective using this curriculum and applying the techniques Paul had taught us.

I found it invigorating to be back in education and was totally consumed with running Success. It felt like ages since I'd been in politics. It was quite jarring, therefore, when I got a message one day that someone from my old life had called: Randi Weingarten.

7

THE CATCH

2007

Randi Weingarten had lambasted me for my union contract hearings, boasted of helping Scott Stringer defeat me, and predicted I'd fail at schooling. Why on earth would she be calling me? When I returned her call, she explained that she wanted Success to adopt the "thin contract" that Bloomberg and Klein had sought. While she'd refused to accept such a contract for the district schools, getting Success to adopt one would be a victory for the UFT since our teachers weren't unionized.

While a thin contract was certainly preferable to a fat one, it would still severely constrain our ability to manage our teachers and create a negative management/union oppositional dynamic. Weingarten asked me, however, whether there was anything she could offer to induce me to change my mind, so I told her I'd give that some thought and get back to her. I did so the following day. I'd agree to a thin contract, I said, if she'd agree to one for the district schools as well. After all, if this thin contract was so great, why limit it to Success?

I never heard from her again.

While my teaching abilities were improving, I was still a relative beginner, so I tried to help our teachers improve by learning from one another. We compared our results and tried to figure out what the teachers with the strongest results were doing differently. Some didn't like their results being made public but I explained that improving their teaching was more important. Imagine if one heart surgeon at a hospital had a dramatically lower mortality rate than the others.

Wouldn't you want to tell this to the surgeons so they could figure out what the one with the better results was doing differently? And while it's true that the lives of our children weren't at risk, their futures were.

Some teachers also felt that since they were all doing pretty well, there was no point in obsessing over slight differences. Aiming for pretty good, however, is a recipe for mediocrity, as explained in the wonderful Charles Osgood poem "Pretty Good," which I posted in the teachers' work room. Every teacher, even the best ones, should strive to become better. One reason Michael Jordan had such an impact on the Chicago Bulls was that his teammates saw how hard he practiced and they figured that if the best basketball player in the world was still striving to improve, so should they.

Many people who find themselves in demanding jobs think they'd be happier in a more relaxed environments but I find that once people adjust to a demanding fast-paced environment, they find it exciting and fulfilling. Conversely, many people who get undemanding jobs are initially very happy because they are comfortable and relaxed but quickly become bored. Hard work is like swimming in the ocean; the cold is unpleasant at first but once your body adjusts to the temperature, it's fun. The trick is having the courage to wade in.

So we all worked together to improve. We regularly observed one another's classrooms and then discussed what we'd seen: the great moves we'd witnessed, the mistakes we'd observed, our suggestions for improvement. We even videotaped ourselves and watched key parts of the lesson. Once my teachers got past their initial discomfort, they found that they enjoyed learning from one another since teaching can be an isolating profession when each teacher is siloed away in her own classroom.

As the year progressed, I thought about which teachers we would rehire for the following year. I'd already let two go and there were four others I felt were weak. I asked Joel Greenblatt and John Petry for advice. John responded by citing Jack Welch's principle that employees generally fall into three categories: 20 percent who are superstars; 70 percent who can be successful with sufficient guidance and training; and 10 percent who aren't right for the company and must be let go. Since we'd already let a few teachers go, John reasoned, the four I was worried about probably fell into the 70 percent category so we should keep them and try to help them improve.

This is just one of the many times that John and Joel gave me wise advice that reflected their considerable business experience. I nonetheless worried about how to square that advice with establishing a culture of excellence at Success. Once you've established such a culture, you can keep on a few weak employees you're trying to improve without others concluding that mediocrity is considered acceptable. These four weak teachers, however, were a third of our staff so I sought to steer a middle course. I told these teachers that while we'd help them to improve, they'd only be rehired if they demonstrated they were truly committed to improving and had made sufficient progress by year's end. All but one did.

However, since we were expanding, we still needed to hire several new teachers and nobody had time to do this as we were all focusing on running the school. Fortunately, Kristina Exline, who'd been on my city council staff and was now working for a women's rights organization, called me to tell me that she was finding her new job boring and wanted to work for me again. I was delighted to have her back and put her in charge of recruiting. With her addition, every one of the five former employees I'd asked to join me at Success had agreed to do so. While I am a notoriously demanding boss, many people find the stress and long hours are outweighed by the excitement and feeling of satisfaction that comes from tackling important projects with terrific intensity. I'm very grateful for their sacrifices.

In addition to teachers, we also needed to hire a dean to help us with the many challenges our students were facing. One student's father had been murdered and his mother had serious psychiatric issues, so we had to get his uncle to take him in. Another student went around saying that he wanted to kill people. We needed a dean who could ensure these problems were handled properly so our teachers could focus on the needs of all of our students. Running an urban school sometimes requires you to balance your obligations to students who are in crisis with those who aren't but whose futures nonetheless depend on getting a strong education.

Finding a strong dean wasn't easy because we needed someone who cared deeply about kids and understood the challenges they were facing but also shared our belief in setting high standards for student behavior. None of the candidates impressed me until I met Khari Shabazz, a former police officer who, tired of arresting young men whose problems arose from societal ills, had quit the force and begun working for an after-school program. Impressed by the financial sacrifice he'd made to help kids and also by his intellect and strong but loving demeanor, I decided to hire him.

While we were making progress in many areas, dealing with DOE proved challenging. It kept losing our requests to provide services to students with learning disabilities. When we protested, DOE then managed to "find" them but said it was now too late to approve the services before the year's end. I was apoplectic and vowed to prevent this from happening the following year.

Another problem was our fluorescent lights, which were constantly going out. It turned out that DOE was purchasing bulbs that lasted only 5,000 hours, so we replaced them with 30,000-hour bulbs. This led to bitter complaints from the handymen who'd been earning copious amounts of overtime pay by constantly replacing DOE's short-lived bulbs. Welcome to public education.

But apart from such frustrations, I felt the year had gone reasonably well. There was a good sense of esprit de corps, the students seemed to be learning, and we were making progress with our challenging students and their parents. Losing our principal turned out to be a blessing in disguise because it forced me to learn more about instruction and while I certainly wasn't the most experienced educator, the teachers saw that I was there when they arrived and when they left and got emails from me at all hours of the night and on weekends. This hard work showed them how important I considered our mission and gave me the moral authority to ask them to take on the herculean task of getting our disadvantaged children to match the achievements of privileged, suburban children who attended schools with twice our resources.

But there was a catch. John, Joel, and I had gotten it into our heads that we were going to have a material impact upon public education in New York City during our lifetimes by opening forty schools within a decade. While other charter schools had "replicated" before, they usually didn't seek permission to do so until they could demonstrate success of their first school with standardized test scores, which our students wouldn't take until third grade, and even then, most only sought to open one more school. If we took that approach, it would take decades to open forty schools. I therefore decided to apply immediately and to ask for three schools rather than one. People told me I was nuts, that it would be a miracle if I got permission to open even one additional school. Moreover, they didn't know about my run-in with SED over our name change. What were the odds that regulators who didn't even want to let me add a word to the name of our first school would allow me to open three new ones?

8

SONGS OF A REFUGEE

1882–1964

My great-grandmother Annie Einhorn was born in 1882 in a portion of the Austro-Hungarian Empire that is now Poland. Having already lost her father, she was orphaned when her mother died from a rabid dog bite. At the age of fourteen, she was sent by her brother to America to become an indentured servant. Upon the completion of her servitude, she married Samuel Ehrenreich, a Polish immigrant, who on May 7, 1898, had arrived at Ellis Island where clerks had dutifully recorded, in accordance with the law of the time, that he was not deformed, crippled, or a polygamist.

Samuel pressed clothes twelve hours a day in a factory while Annie opened up a small grocery store. She was illiterate and innumerate but one of her customers taught her how to add, multiply, and subtract so she could make change. In 1905, she gave birth to a girl, my grandmother Frances.

Frances became a public school teacher and, in 1931, married Emil Moskowitz, a child of emigrants from Budapest and, like his father, a tailor. In 1935, Frances gave birth to a boy, my father, Martin. My father would help out at his father's store on weekends. During a period when suppliers were insisting that Emil pay cash due to his financial difficulties, my father once had to deliver $3,000 in cash to a supplier. Understanding full well the disaster that would befall his family if he lost this enormous sum, my father breathed a sigh of relief when he arrived at his destination.

One day, a boy named Clement Finn called my father a "dirty Jew," so my father slugged him as hard as he could and was sent to the principal's office to explain himself, which he happily did: Clement had it coming. Alas, that view did not prevail so my father was promptly shipped off to the Pleasantville Cottage School for troubled children.

My father discovered he had a natural affinity for mathematics that others didn't seem to share. One day, during a lesson on how to tell time, my father raised his hand and observed that in theory, you didn't need the minute hand to tell time if you looked closely at the hour hand. His teacher told him to shut up.

In third grade, my father noticed a pattern in the multiplication tables. If you took a square (e.g., 8 × 8), and then you made one number lower and one higher (e.g., 7 × 9), the result (63) was one less than the square (64). He saw this was true for all the squares up through 10 x 10, which was where the multiplication table ended, so he asked his teacher whether that would continue on forever. Again, he was told to shut up.

Later, my father figured out that the paper cone in which the egg cream sodas he liked were served would hold one third as much as a cylinder of the same dimensions. He tried to explain this to the store's proprietor but, to my father's surprise, the proprietor couldn't grasp this concept.

Sensing my father's gift, my grandmother had him take the admissions test for Stuyvesant, the city's best public high school, and he passed. There, he was pleased to find others who shared his enthusiasm for math, although he was upset to learn in his senior year that he wouldn't be allowed to take calculus. He was told that only students on the math team could take calculus, and when he offered to join the team, that he couldn't join the math team just because he wanted to take calculus. He resorted to taking it at night school instead.

After graduating, my father enrolled in an engineering program at Cooper Union but decided after two years to transfer to Brooklyn College to study pure math. My grandmother opposed this plan as impractical and wouldn't speak with my father for six months when he went ahead with it anyway. At first, my father struggled, but in the end, he did well enough to gain admission to a graduate program at the University of California, Berkeley. There, he studied with Gerhard Paul Hochschild, one of the many Jewish refugees from Germany who became leading lights in the fields of mathematics and physics. Spurred on by the same fascination that had led him to discover the relationship between the volume of a cone and a cylinder, my father came to study Lie groups, a field that lies at the intersection of geometry and algebra.

After a year of studies, my father returned to New York City for the summer. When it ended, he drove back to Berkeley with my mother, whom he'd met at an Israeli folk dance. While my father didn't exactly conform to my mother's romantic ideals, their cross-country journey allowed her ample time to discover virtues in him not apparent at first sight. They began dating and, in 1959 got married. As the ceremony took place in California, their families couldn't attend. The only guests were two witnesses, the spread for the reception was half a wedding cake, and the honeymoon was a day trip to Sausalito.

My parents decided to move to Israel where my father got a position at Hebrew University. While sad to see his daughter move so far away, my grandfather Chaim nonetheless "bless[ed] the hour when [she'd] set foot in the promised land." My parents ultimately decided, however, that the promised land wasn't their cup of tea and returned to the United States, but first, they took a boat to Naples and hitchhiked their way throughout Italy and France. By spending just $2.50 per day four years after publication of the budget travel guide *Europe on $5 a Day*, they managed to make $600 they'd saved last four months. My mother was so inspired by the sites she saw that she later decided to study art history.

In 1961, my grandfather Chaim learned he had cancer, the same illness that had felled his beloved Sascha. It was now time, he realized, to fulfill the vow he'd made to publish the Yiddish poems he'd written. Fittingly enough, he funded the publication of his book, *Songs of a Refugee*, with reparations payments he'd received from Germany.

He explained in the book's preface that he wished to depict "the deep sorrow of the Jewish refugees who were persecuted along the rocky, thorn-strewn road as they were chased from border to border" and "the joy of those who finally found refuge . . . in the free America."

His poems were filled with gratitude for the country that had welcomed him:

Now, after wandering so long,
over many lands and oceans wide,
I sing a song of praise to you,
the land that gave me a new home.

Now, no one is excluded,
that is what makes you truly great.
All the children of God's world
are welcomed at your gates.

So accept the gratitude
of a simple immigrant.
You provided us with a safe haven.
Be blessed, you precious land!

His poems also expressed his love for his adoptive city:

I'm enchanted by your long, broad streets,
your towering palaces of stone and steel,
the multicolored din of your Times Square
which calls to joy and life both night and day.

Your Greenwich Village artists, bearded poets,
oft laughed at, mocked, yet so creative;
by day, they paint out in the sunshine
by night, sip wine in local pubs.

But while grateful that he'd managed to find refuge in America, he lamented that doing so had opened up an unbridgeable cultural divide with his children, who couldn't understand the Yiddish language he so loved:

I have two daughters, beauteous as roses,
Magically charming as all can see.
But my ancestry is foreign to them.
Who can share my deep sorrow with me?

I have a cupboard full of books
But nobody to leave those tomes.
The attic and the cellar will read them
when my final hour comes.

And he was still haunted by the horror the Nazis had inflicted on his family and his people:

The bloody wounds have not yet healed;
the killings, the torture, fresh in my memory.

I can still hear the moans of the martyrs—
their lives cut short by the German hordes.

Not even one generation has passed,
yet so much of what happened is being forgotten.

Remember! Remember to the ends of time
how we were herded into the fire and gas.

Chaim's life had not been easy. His wife had died of cancer, his mother had been murdered in a gas chamber at Treblinka, and he'd struggled financially in New York. But his final poem suggests that, notwithstanding the losses he'd suffered and the anger he felt, he'd found a measure of comfort in his adoptive home and the knowledge that his descendants would enjoy a happier life in the land to which he'd delivered them. This poem, titled "Sunset" and written by Chaim at a Jewish country retreat after his cancer diagnosis, suggests he was at peace with his own imminent mortality:

The fragrance of flowers
spreads through the air
engulfing all
as sunset approaches.

A narrow cloud, now flaming red,
shimmers in the sky;
reflected in all its beauty
in the waters of the lake.

Rowboats glide quietly toward the shore.
Couples disembark and bring along
the blessing of the cool waters
in their song of praise to the sunset.

Six months after Chaim wrote this poem, my mother gave birth to a boy she named Andre and traveled back to New York so my grandfather could meet the first of his descendants to be born in America. By December 1962, it was clear the end was near so she returned again. Despite the terrible pain he was enduring, Chaim smiled to see his grandson one more time. Days later, he died.

A little more than a year later, on March 4, 1964, my mother gave birth to the second of Chaim's descendents to be born in America, a girl to whom she gave the name Eva Sarah Moskowitz.

9

GET US THE DAMN SPACE!

2007–2008

Normally a charter school seeking to replicate sought permission from its authorizer, which in our case was the State Education Department (SED), but given our history with them, we instead applied to the State University of New York (SUNY), which also had the power to grant charters. I knew the head of its charter school committee, Ed Cox, a prominent lawyer who liked the work I'd done on the city council, so I arranged for Joel, John, and me to meet with him and he encouraged us to apply.

In the summer of 2007, we submitted our applications which, due to SUNY's onerous requirements, consisted of 45,000 pages contained in ninety binders. SUNY got back to us on August 16. They had one big concern: if we couldn't open because the city didn't give us space for our schools, SUNY would have wasted three charters, so they asked us to promise we'd rent space if necessary. I refused. To open enough schools to make a serious dent in the city's educational crisis, our schools had to operate, aside from start-up costs, on public funding. Renting space would make that impossible since it would add around $2 million in costs annually per school. If we reached our goal of opening forty schools, we'd have to raise $80 million annually. Making ourselves dependent on such a huge stream of philanthropy would be fiscally irresponsible; one bad year and we'd be thrown into a financial crisis, forced to choose between eviction and laying off teachers. Besides, paying rent would be a waste of money since plenty of district schools in Harlem were half empty. Parents there preferred to send their children to parochial schools, charter schools, or district schools in other neighborhoods. At one nearby school, PS 241, enrollment had fallen from 918 students to 347 in just three years.

But SUNY was adamant. If we wouldn't promise to rent space, said SUNY, then we had to get DOE to commit to giving us space and while I was confident DOE would do so once we ran our lottery, they were too lethargic and cautious to make that commitment up front. It was a classic catch-22: DOE wouldn't give us space until SUNY gave us charters and SUNY wouldn't give us charters until DOE gave us space.

While this was playing out, our first school reopened on August 20. This time we actually had a plan for making sure our students ended up with the right teacher, but we had to refight some battles. The piercing custodial bell had returned because, as we soon learned, the custodian wouldn't answer his walkie-talkie due to a feud with PS 149's principal over a romance gone bad between their respective staff members. Another issue was syrup, as I wrote Joel and John:

> Last year I outlawed syrup for breakfast. Five-, six-, and seven-year-olds can manage to get the entire school sticky. First day of school syrup was served. Food service managers told me that I did not have the authority to outlaw syrup and that only the US Secretary of Agriculture could make that decision.
>
> Who knew?

In addition, our special education teachers were being paid late by DOE, so we had to lend them money. DOE also took forever to hang a curtain in our auditorium because curtain hangers were unionized and in short supply. Despite having held more than one hundred hearings as chair of the Education Committee, I'd never heard of this union.

But some things were going more smoothly. Khari, our new dean, was a big hit with the kids since he had a kind but authoritative manner and an orange BMW motorcycle he'd let them sit on if they were well behaved. He was also deeply compassionate. When Khari learned that one of our students was angry and defiant because he missed his father, who was imprisoned many hours away from the city, Khari offered to bring this child to visit his father. The boy's attitude immediately changed; he started smiling and talked about showing his father his writing and teaching him how to play chess. It was heartwarming and yet at the same time heartbreaking to see how this boy longed for something most children could take for granted. After the visit, the boy's mother thanked Khari profusely, saying that she "didn't know people could care that much."

Khari was also more comfortable than many of our teachers with speaking frankly to our parents. He wasn't afraid that telling them they had a moral obligation to check their children's homework would be perceived as a cultural judgment. He also nimbly handled the disputes that sometimes arose between parents. One day at dismissal, several mothers of our students began assaulting a student's father with bags chock-full of library books because they believed he was two-timing his wife. Unfortunately, when we'd instituted our policy requiring our parents to read their children books, we hadn't anticipated the risk that they'd be weaponized. I tried to intercede but, at five foot two, was ill suited to the task. Fortunately, Khari stepped in and calmed everyone down.

We also systematized certain practices like regularly checking the cleanliness of our school. Every time we found a deficiency, we took a picture of it that we forwarded to the custodian. Another innovation was requiring teachers to call parents regularly. Imagine this. You're making dinner and the phone rings. It's your son's first-grade teacher. "Oh no," you think, "what has he done now?" The teacher says, "I wanted to let you know that Taj wrote a wonderful story today about our trip to the circus. You should ask him to show it to you." Buttering you up for the bad news, you think, but the bad news never comes. The sole purpose of these calls was to build a positive and productive parent-teacher relationship, something that's much harder to do if a teacher's first interaction with a parent is negative.

We also encouraged our teachers to talk less. In math, teachers were supposed to provide only eight to ten minutes per day of "direct instruction," meaning leading discussion from the front of the room. Children learn better by doing: reading books, solving math problems, writing, having conversations. In theory, our teachers already believed this: ask them about progressive education, and they'd happily talk your ears off about its virtues; but put them in front of children and many would drone on endlessly. We therefore made them set kitchen timers so they'd know when to stop.

Meanwhile, I finally managed to break the DOE/SUNY logjam. DOE gave us a letter regarding their intention to give us space that was firm enough to satisfy SUNY but vague enough that DOE felt it could back out if necessary. Thus, on October 26, 2007, with everyone's bureaucratic posterior now covered, SUNY approved our applications to open three new schools, and we soon received two grants to help open them: another $1 million from Don and Doris Fisher and $660,000 from the Walton Family Foundation.

But SED, which had the power to veto SUNY's approval, wasn't satisfied with the 45,000 pages of applications we'd submitted. It made seventy-three requests for more information on such educationally critical matters as "the process for completing bank reconciliations" and putting column headings on the second page of a chart that spanned two pages. We responded to these requests but SED rejected our applications anyway, citing, among other things, our lack of standardized test scores. In this regard, SED was oblivious to the practical reality facing families in Harlem. Sure, they would have loved to send their children to a school with a proven track record of success, but there weren't any in Harlem. Rather, they had to choose between schools that had proven themselves to be lousy and our schools, which, while unproven, at least might be good. Many families believed we were a better bet and they were in a far better position to make that judgment than bureaucrats hundreds of miles away in Albany.

Fortunately, however, SUNY had the power to override SED's veto and did. Now I needed to get co-location sites from DOE. Although

I'd identified six buildings that each had more than 450 empty seats available, DOE hadn't yet committed to specific sites, so I asked Joel Greenblatt to call Chancellor Klein and be firm with him:

> His lieutenants treat us as if we're asking for special favors. We're taking kids off his hands in a neighborhood that has been an education disaster. He needs to tell his lieutenants to be thankful and get us the damn space!

In late February, DOE announced a site for one of our schools, PS 123, but opposition soon emerged. Council Member Inez Dickens claimed there was "a noticeable shortage of space at PS 123," although she failed to explain how that assertion could be squared with the fact that PS 123 had only 540 students in a facility built for 1,082 students. Six days later, one of her aides boasted at a community board meeting that Dickens had "successfully fought twice to make sure charters don't go into public school space" and would "fight again."

Interestingly, a teacher at PS 123 who attended this meeting conceded that there actually was space at PS 123, but complained that this was because charter schools were "robbing the school system of children," as if these children were chattel who belonged to the schools rather than human beings entitled to receive a good education. Another teacher claimed Success wasn't needed because "We have school choices. Public schools. You can go to any one in the city." Obviously, that just wasn't true. The good schools were generally located in affluent neighborhoods and had few spots available for students who lived elsewhere. Yet another teacher claimed that charter schools only admitted students who'd scored high on state tests and "leave us with the 1s and 2s." In fact, the law requires charters to admit students by lottery, and students weren't tested until years after we admitted them. Finally, the head of the community board said to a white DOE representative, "You are creating opportunities for your cousins coming in. This stuff is not for us."

A hearing on our co-location was scheduled for April 1 and I feared the UFT would gin up opposition by using scare tactics with PS 123's parents. While Bloomberg supported co-location, there was a limit to how much political capital he could afford to spend on us. We needed to make a strong showing at the hearing so that both the Bloomberg administration and the press would understand the broad support for charter schools in Harlem. The problem was that, while we'd received three thousand applications, we hadn't run our lottery yet, and parents who didn't know if they'd be able to send their children to our school even if it opened didn't have enough motivation to come out and support us. So we had to rely upon the families of our existing students. Anticipating this day, I'd told them at the outset that if they sent their children to our schools, they should be willing to fight to give other families that opportunity. Most agreed, as one parent, Kyesha Bennett, explained in an op-ed piece:

> My mom gave me an early lesson in school choice. She got me into PS 87 on the Upper West Side. I was 9, and I had to get myself and my 6-year-old sister to a school 2 miles away by public bus. Maybe it sounds crazy for kids to travel alone in New York City, but my mom believed that going to my "zoned school" was more dangerous for my future.
>
> Flash forward 22 years. I needed to find a school for my son. I was shocked to find my options in Harlem were almost identical to my mother's. I heard about a charter school in the neighborhood. I prayed for him to get in and he did.
>
> All parents should have good school choices. That's why I've joined with other mothers and fathers to found a group called Harlem Parents United.

The day before the hearing at PS 123, I was explaining to a *Daily News* reporter why co-location was controversial and I said that the endless conflicts in the Middle East showed that "dividing land ain't pretty." I thought I'd made my comment off the record, but the reporter apparently understood otherwise because she published an article stating that I was "gearing up for . . . a 'Middle East war' over classroom space." This was bad. It sounded like I was bringing Zionism to Harlem with African Americans as stand-ins for the Palestinians. I'd played right into our opponents' hands. I apologized publicly and also privately to John and Joel.

On the day of the hearing, the auditorium was packed with parents from both Success and PS 123 and it quickly became apparent that, just as I'd feared, the UFT had given parents false information. One parent claimed that if our co-location was approved, PS 123's class size would increase to thirty-five students. This wasn't true. Space wasn't what determined class size at PS 123. If it did, then PS 123 would have had nine students per class since it had fifty-eight full-size classrooms for its 538 students. What determined PS 123's class size was the number of teachers it could afford to hire, which wouldn't change.

Our parents sought to counter the UFT's efforts to portray us as outsiders:

> "I have four children. Two in public school, one in charter, and one on her way to charter . . . We're no better than anybody else sitting here. We all bleed the same. We all shop at the same supermarkets, we all go to the same hospitals when we're ill. [W]hy can't we cohabitate here . . . ?"
>
> "I am a teacher at public schools with the Department of Ed . . . [T]here [is] space available. We're just trying to share."
>
> "People keep trying to [make race] an issue—if you're black, if you're white, if you're purple. I wouldn't care if the teacher was green! My baby can read!"

Perhaps our efforts to show support for co-location were successful since DOE soon announced two additional co-location sites for us: PS 7 (which had 425 students in a building that could hold 867) and PS 101 (which had 612 students in a building that could hold 1,190).

On April 16, we held our lottery at the Mount Olivet Baptist Church. Thirty-six hundred applications were submitted for six hundred seats including more than 40 percent of the District 5 children eligible for kindergarten. Among them was my son Dillon. Like many other families in Harlem, Eric and I didn't really have any good alternatives and were just praying he'd get in.

We encouraged the press to attend the lottery, which some people felt was cruel since it would make a public spectacle of an event that would be very painful for those families who walked away empty-handed. I was sympathetic, particularly since I too had a child in the lottery, but I felt it was important for the public to understand the desperation of families in poor communities for better educational opportunities since it would help build pressure for change. Thankfully, Dillon got in, so Eric and I breathed a sigh of relief, but for most families, the lottery was incredibly painful.

As I'd predicted, now that we'd identified the six hundred children who would be deprived of the opportunity to attend our schools if our co-locations were nixed, opposition to our co-locations quickly died down. Now we just needed to open our three new schools simultaneously by August.

10

I HOPE MY BOARD FIRES ME

2008

There are two tragedies in life, said Oscar Wilde: not getting what you want and getting what you want. I was experiencing the latter. Having received permission to open three new schools, I now had to find three new principals, hire and train eighty new teachers, renovate three new facilities, and prepare to educate six hundred new students, all in a few months' time and while I was still figuring out how to run my first school. I was now even more grateful that Kristina Exline had returned since she possessed the naive optimism of the character Rose shooting the rapids in *The African Queen*. Impossible tasks that might cause others to crumble were just another exciting adventure for her. Fortunately, her fearlessness also meant she made quick decisions. When we got a good résumé, we'd call the candidate that day to schedule an interview and, if it went well, make an offer almost immediately.

Given that most of the teachers we were hiring were inexperienced, we implemented techniques and systems they could learn quickly. We taught them that when they read to students, they should periodically suggest questions students should be thinking about, such as whether a character is being truthful or why a character is angry. Like Paul's techniques, these questions encouraged children to engage in "active listening"—thinking about the text rather than just passively taking in its literal meaning. Another technique was "turn and talk": a teacher asks a question, each student discusses it with the student next to him, then the teacher picks one student to repeat what his partner said. This gives students more opportunities to actively participate.

We also developed hand signals to keep class running smoothly. Instead of asking students to raise their hands if they agreed with a classmate and then if they disagreed, we'd ask all of our students to simultaneously give a thumbs-up or thumbs-down. We also had signals for needing a drink of water, a tissue, or permission to use the bathroom, which enabled a teacher to simply point to a child to grant permission. Without these signals, teachers would call on students expecting them to participate and would end up with bathroom requests, which would disrupt the short attention span of young children. Techniques like these may seem unimportant, but if they save 5 percent of your class time, that's the equivalent to adding two weeks to the school year.

But to be effective, the entire school has to adopt these techniques. While college students can quickly accommodate themselves to different teaching styles, young children find it hard to learn even one set of rules and expectations, much less several. Occasionally I'll read about a teacher at some school who refuses to teach contrary to her deeply held pedagogical beliefs, which are different from those of the school's principal. That just doesn't work. If a teacher doesn't agree with her school's pedagogical philosophy, she should find a school whose philosophy she does agree with, not create chaos by marching to the beat of her own drum. It's not that the school's principal is necessarily right, it's that a team can't have multiple quarterbacks calling different plays.

Of course, our rookies didn't become master teachers overnight. Because they lacked confidence in their students to get the right answer, they'd often steer students to it in a rather ham-handed fashion. It takes time before teachers have the confidence to let students think for themselves. Nonetheless, an inexperienced teacher could get pretty good, pretty quickly by using the techniques we taught them.

By July 28, the first day of training, we'd managed to hire all but two of the eighty staff members we needed, and Teach for America

(TFA) asked us to hire a teacher who'd gotten into trouble at a district school. He and his fourth-grade class had come up with an idea for a silly little film in which the students would briefly pretend they'd turned into animals—cows, monkeys, whatever they wanted. The kids enjoyed it but one parent complained that it fed into a racist stereotype of urban kids acting like animals, which prompted the school's principal to bring the teacher up on disciplinary charges for failing to get her permission for this project.

Rather than fight the charges, the teacher asked TFA to reassign him. We interviewed him and found him to be bright, idealistic, and well intentioned. Problem solved, right? Not quite. The teacher found out that DOE still wanted to proceed with disciplinary charges. Fortunately, I was able to get DOE to back off but I still found it incredible that DOE would have bothered with this teacher once he'd resigned.

Just weeks earlier, an article had appeared in the *Daily News* about DOE's efforts to dismiss a teacher who'd bombarded a fifteen-year-old student with lovesick emails but whom an arbitrator had declared fit for duty. Fortunately, DOE had gotten a judge to reverse the arbitrator's decision, but the cost of proceedings like these was huge, as the journalist Steve Brill would soon reveal in a *New Yorker* exposé. Teachers whom DOE was seeking to terminate were consigned to what was known as the Rubber Room, a sort of purgatory for teachers. One termination proceeding on which Brill reported was expected to last forty days, eight times as long as the average criminal trial. Moreover, it made little sense to have an arbitrator second-guess a principal's determination regarding a teacher's competence since the arbitrator couldn't directly observe instruction. This was underscored in one hearing by the UFT's use of a photograph showing the teacher at issue in a well-organized classroom with a model lesson plan written on the blackboard: the picture, it turned out, had been staged with the help of the UFT chapter leader.

Moreover, since the UFT could object to the appointment of an arbitrator whom it believed had been too harsh in a prior case, arbitrators would often "split the baby" by fining teachers and throwing them back into the classroom. The process typically cost hundreds of thousands of dollars when you took into account not only the cost of the lawyers and the arbitrator but also the salary of the teacher, who was paid for sitting in the Rubber Room while this charade played out.

Yet while DOE seemed to have its hands full with teachers who were incompetent or romantically obsessed with their students, they'd somehow found the time to persecute a promising young teacher with a 3.9 average from an Ivy League university who'd made at worst a rookie mistake *and had already resigned*. This episode illustrates how a system of charter schools can actually be fairer to teachers than a government educational monopoly. With charters, a teacher fired at one school can usually get a second chance at another. Where district schools are the only game in town, however, termination can end a teacher's career and, because this is such a severe consequence, it leads to demands for "due process" protections that end up being expensive, time consuming, and ineffective.

Happily, with the addition of this young man, we now had a complete set of teachers and all we had to do now was teach them to teach. DOE trained its teachers for one day before they started; we did so for nearly a month before school began as well as a dozen additional days throughout the year and every Wednesday afternoon. Unlike teachers' colleges, at which professors with little classroom experience give lectures on pedagogical theory, our principals, deans, and strongest teachers delivered our training and focused on concrete skills: how to prepare for class, handle discipline issues, teach children with special needs, talk to difficult parents, use the classroom techniques Paul had taught us, and implement components of THINK literacy, the curriculum we'd built.

We also trained our teachers in a new mathematics curriculum we were rolling out called TERC, which was "constructivist," meaning that it focused on conceptual learning and nurturing students' own intuitive understanding of math. The comedian/mathematician Tom Lehrer once said that the point of what was then called "new math" is "to understand what you are doing rather than get the right answer." There's an element of truth to this. In the age of the smartphone, facility with long division is less important than understanding mathematical principles and developing problem-solving skills.

But there was one problem: SED told us its Board of Regents had to vote on any "physical change to the charter documents," meaning any change to a single word of these documents although they were hundreds of pages long and included such minute details as the exact length of every class. Moreover, since the regents weren't going to meet for several months, we'd have to delay adopting this curriculum for a year. SED's attitude was: not our problem. They cared about regulatory processes, not a silly little thing like making sure kids had a good math curriculum. SED even threatened to put us on probation if we went forward with TERC, which worried John Petry. "I could see the story of 'Moskowitz School on Probation,'" he wrote Joel and me. I responded, "Yes, but I can't run a school where [every single change] requires a charter revision." So we went full speed ahead, damn the torpedoes, and just as with our school name, SED blinked.

Another challenge was renovating our facilities in just a few weeks. As with so many things, half the work was fighting people trying to stop us. On August 15, the head of the local painters' union showed up to harass our painters. For decades, the painters' union had been controlled by the Lucchese crime family, and more recently, one of its top officials was prosecuted for kickbacks and thefts totaling nearly $750,000. Our painters were quite intimidated so they decided to do the remainder of their work on nights and weekends to avoid further confrontations.

Our final challenge was arranging the schedule for sharing common space such as the gym, cafeteria, and play yard. PS 123's principal refused to discuss this with us until her school opened, which was two weeks after ours did. This meant we'd have to start with one schedule and then change it two weeks later. I begged Deputy Chancellor Garth Harries to intervene because changing our schedule two weeks into the school year would be quite disruptive. He replied: "Adjusting schedules midstream happens all the time." I responded, "If the day ever comes when I think something is okay simply because district schools do it, I hope my board fires me." Excellence is the accumulation of hundreds of minute decisions; it is execution at the most granular level. Once you accept the idea that you should give in to things that make no sense because other people do those things and you want to appear reasonable, you are on a path towards mediocrity. To achieve excellence, one must fight such compromises with every fiber of one's being.

But despite my fervent protests, I lost this battle. Thus, after school started, we had to rip up our schedule, tell our part-time art and gym teachers they now had different schedules, and change routines on our kindergartners and first-graders. It galls me to this day.

As opening day approached, problems seemed to multiply: the kindergarten tables we'd bought were delivered without legs; our telephone service was knocked out by a car accident; a first-grade teacher quit the day before school started; we inadvertently overenrolled kindergartners; one of our play yards was littered with bullets. But somehow it all came together and on August 25, we opened the doors at our four schools to a total of nine hundred kids.

11

CATNIP JELLY

1966–1976

In 1966, my father accepted a position at Columbia University. New York City had changed considerably in the decade my parents had been away. The manufacturing jobs in which the city's immigrants had long toiled were rapidly being replaced by positions in finance, entertainment, and tourism. Wall Street generated enormous employment both directly and by feeding the city's law and accounting firms. So too did the television networks, which had their headquarters and produced much of their programming in New York. These industries thrived in large measure because of the city's unparalleled ability to attract talent. Those with ambition and ability flocked to the city to work with, and compete against, the very best, and success begat success, as exemplified by the Yankees's sixteen World Series championships over a twenty-six-year stretch.

For my parents, however, New York was simply home: the city where they'd grown up, still had family, and felt most comfortable. They rented an apartment near Columbia's campus and formed a babysitting cooperative with some young professors and graduate students in the neighborhood. When my father took his turn, he'd have us run races around our block while he sat on the stoop of our apartment building working on math problems. My mother often read to us; I fondly recall *The Secret Garden; Heidi; The Call of the Wild;* and especially *The Wise Men of Chelm and Their Merry Tales,* a book of Jewish folktales that never failed to send Andre and me into hysterics. My great-aunt Minna also helped out, sometimes by taking me to the movies. I loved *The Sound of Music* so much that I prevailed upon her to sit through three showings with me.

My mother enrolled in a graduate degree program in art history at New York University and, as the program's rigorous qualifying exams approached, she had less time for Andre and me. "Once you get the PhD," I asked her one day, "then will you be a real mommy?" Naturally, she was quite hurt, which I feel bad about. She was very much a "real mommy," who ate dinner with us and read us bedtime stories every night, but I was so terribly fond of her that I craved even more attention.

In the summer, my father would drive us to California so he could collaborate with Professor Hochschild. As we were still quite poor, we traveled frugally. For breakfast, we'd bring our own cereal to thruway malls and surreptitiously grab fistfuls of coffee creamers that we'd use as milk. We often slept outside in sleeping bags, on one occasion on a nest of biting ants that soon let us know we weren't welcome. We owned a Plymouth Valiant, a car model that lasted forever, so that's how long my dad decided to keep it. When we ran into hot weather, my parents would buy bags of ice to put on our laps, which they told us was "air-conditioning." One day, a doorman of a building in front of which we'd stopped futilely attempted to open a door on our Valiant that had long ago ceased working, and was bewildered when Andre and I proceeded to roll down the window and clamber out.

In the fall of 1968, I began attending a wonderful integrated nursery school in the middle of Morningside Park. During art class, we'd wear smocks and berets and pretend we were French painters. The following year, I went to my District 5–zoned school, PS 36, where my brother and I were virtually the only white children.

As I look back, I am struck by how early I developed the interests that would occupy me for the rest of my life. I loved pretending to teach. I'd set my stuffed animals in front of an easel blackboard and instruct them in addition and subtraction, which my father had taught me, or read from picture books whose words I'd memorized.

I eventually managed to convince a few children in our apartment building to join my class but they weren't very well behaved, so I had to reprimand them for their lack of effort. They complained that I was too strict and, sadly, soon abandoned their studies for jacks and jump rope. My stuffed animals, however, stoically persisted in their studies.

Trips to historic sites such as Colonial Williamsburg, Mystic Seaport, the Paul Revere house in Boston, and Independence Hall in Philadelphia spurred my interest in history. So too did a trip to Italy one summer on which we saw ancient churches, medieval castles, and villages which led me to imagine what life was like back then.

I found it fascinating to listen to my great-grandmother Annie, who lived until the age of ninety-six and seemed impossibly old to me: her skin was as wrinkled as a prune; her hands were bony and gripped me strongly; and her ears, which faced forward like those of a monkey, became comically large with age. She delighted in telling me how different things had been in her time: how the subway had cost just a nickel and doctors made house calls in horse-drawn carriages. It seemed incredible to me that someone who had lived in an age that was so different could still be alive.

I also developed a strong interest in politics since my family was involved in the political activism of the age. My parents were at Berkeley at the birth of the free speech movement, an early manifestation of student activism, and were at Columbia when students occupied numerous university buildings including the math building, which my father got permission to enter so he could retrieve a manuscript on which he'd been working. My family was caught up in the issues of the day. We attended Vietnam War protests, painted a peace sign on the roof of our apartment building for the planes flying overhead, and regularly discussed politics at dinner.

One summer, I attended Camp Hurley, which had been founded by a social worker with strong leftist political views. Instead of traditional campfire songs, we sang union songs such as "Keep Your Eyes on the Prize," civil rights songs such as "We Are Soldiers in the Army," gospel songs such as "Go Down Moses," and folk songs with political overtones such as "This Land Is Your Land." Sometimes the folk singer Pete Seeger would lead our sing-alongs and every Friday night we'd link arms and sing "We Shall Overcome." The camp's head, Morris Eisenstein, told me I was "bourgeois" but liked my enthusiasm for the antiwar and civil rights protests he encouraged the campers to stage, and I endeared myself to him by engaging in an act of civil disobedience: refusing to obey the camp's curfew the night before a protest so I could finish painting the protest signs.

As a result of these influences, I became quite political. I wrote a letter to President Nixon expressing my opposition to the secret bombing of Cambodia. He responded with a cheery portrait of his family which made me feel he hadn't taken my letter to heart, so I called the White House and let his staff know his response had been wholly unsatisfactory. While my interests in politics, teaching, and history naturally manifested themselves in childish ways at first, they persisted and in time became the focus of my life's work.

Unfortunately, my formal education made little contribution to my intellectual development. I attended PS 36 and every year, we'd spend months relearning math we'd been taught the year before. I was so bored that when I saw that some of my classmates were leaving class to get counseling, I asked to do so as well and talking with the counselor soon became the highlight of my day. As for learning, that happened primarily at home when my parents tutored me. I knew, however, that my classmates were totally dependent on the patently inadequate education they were getting at PS 36, and even at that age, I understood this would put them at a terrible disadvantage later in life, which seemed horribly unfair.

After a couple of years, my parents managed to get Andre and me into PS 6, a school on the Upper East Side, which was a big improvement. I remember one of my teachers, Ms. Goldberg, quite fondly, and am in touch with her to this day. One of my classmates was the musician Lenny Kravitz, whose mother was on a TV show called *The Jeffersons*. Another boy in my class wouldn't stop teasing me, and one day I reached the breaking point so I slugged him as hard as I could and gave him a bloody nose. Fortunately, unlike my father, I wasn't sent away to a school for troubled children.

Since PS 6's lunchroom was noisy and smelly, many students went home for lunch, but Andre and I lived too far from school to do this, so we took the lunches our mother made for us to various places in the neighborhood including Gimbels, a nearby department store where we'd sneak under a table draped with a tablecloth, and the cafeteria at the Metropolitan Museum of Art, where I'd fish coins out of the museum's fountain so I could buy a cookie. One day, a guard reprimanded me for this, which surprised me as I didn't see how it was any different from picking up a coin in the street, but I stopped anyway.

My parents thought our family should spend more time in the countryside—think Walden Pond—so they bought a fixer-upper in an agricultural area of upstate New York. When my father installed heating, he asked me to crawl inside a heating duct, as I was the only one who fit. When I later wrote about this in a school essay, my teacher accused me of lying. I'd already complained to my father about this teacher on several occasions for doing things like eating milk and cookies at her desk each day right before lunch as my classmates and I looked on hungrily. When I told my father about this latest episode, he wrote on a sheet of paper in my notebook "Ms. ___, screw you. Martin Moskowitz." He did it as a gesture, to show me he was on my side. I knew I wasn't actually supposed to give the note to my teacher but one day I got so angry that I did. Yet again, my father was summoned to the principal's office!

Our family would spend the entire summer at our home in upstate New York, but not to engage in leisure activities like swimming or tennis. Instead, we worked. We fixed up the house, picked berries, made jam, baked pies, grew vegetables, and, one year, planted eight thousand saplings. Andre and I also herded a local farmer's cows when they needed to return for milking, earning a quarter each.

My parents were models of industry. Although my mother didn't become a professor until relatively late in life, she ended up publishing seven scholarly books on art history, continuing to do so even after she retired from teaching. When my dad wasn't tutoring us in math or doing his own work, he fixed up the house and, when he finished that, turned the barns on our property into a larger home. Andre and I learned from our parents' example, so they never had to nag us to do our homework.

Our family had found a good balance by splitting our time between the city and the country, but by the early seventies, New York was becoming increasingly dangerous and unlivable. The subways were covered with graffiti, racial tensions were high, the schools were chaotic, and the city's murder rate had doubled since my family's arrival six years earlier. Just blocks from our home, a teenager was shot to death just for laughing at another kid. Andre and I were repeatedly mugged, and our apartment was burglarized so often that our parents told us to knock on the door when we came home to give any burglars who might be inside a chance to flee via the fire escape before we entered. My father eventually embedded glass in concrete on our windowsills and, after a burglar broke down a neighbor's door, reinforced ours with plywood and stuck a chair underneath the doorknob at night.

In 1973, my parents finally threw in the towel and moved our family to our home in upstate New York. Monday through Wednesday, they stayed at a small apartment in Manhattan to do their work, leaving Andre and me in the care of a local woman. At first, I dreaded their weekly departures and felt lonely since the nearest children were miles away. Over time, however, Andre and I adapted. We read, became even more studious, and managed to find ways to amuse ourselves.

In the summer, my mother and I canned vegetables and made jam and I learned how to bake, winning some prizes in local fairs for my efforts. We also kept busy with chores such as fetching unpasteurized milk from a local farmer. Walking back up the dirt road, each of us holding one handle of the heavy milk jug and sporting bright red hair and freckles, we reminded my mother of a Norman Rockwell painting. We often skimmed the milk to make butter or ice cream. At times, however, our enthusiasm outstripped our knowledge. When we told our neighbor that our cat had taken a curious interest in the jelly we'd made from freshly picked mint, she gently let us know that we city folk might not be familiar with the difference between mint and catnip. News of the Moskowitzs's catnip jelly quickly spread among our amused neighbors.

The local school was pretty good and, at our parents' insistence, allowed Andre and me to double up on our science classes because the science teacher was particularly strong. On Saturdays, our parents drove us to Saratoga Springs so Andre could study cello and I could study pottery, which I enjoyed, and ballet, in which I floundered as I apparently lacked both arches and rhythm.

In 1975, my parents decided that we should take a sabbatical to which my father was entitled in Paris. We lived in the Latin Quarter and I attended a neighborhood public school whose students were quite friendly. Every morning, they'd greet me and one another with four alternating kisses on the check. Every new student's arrival provoked another round of kissing, which I found hilarious, and my resulting giggling proved contagious. When the school doors opened, we all streamed in. The door shut behind the last student, so if you were late you had to ring the bell to get in and would receive a severe scolding. Once inside, the clogs that all the girls wore, including me, made a thundering noise on the wooden stairs.

The teachers were very strict and were obsessed with form. There were very precise rules for taking dictation: you had to write on graph paper with a fountain pen, put two lines between your name and the title, put one line between the title and the text, underline certain things once and others twice, etc. . . . When my papers came back covered with red marks because I hadn't followed the rules, my classmates would try to help me understand them, but I'd always get mixed up, which often led to another bout of giggling.

Lunch was a long multicourse affair served on a properly set table with real tablecloths, plates, utensils, and cloth napkins. They even put out pitchers of wine for the children as I learned one day when I accidentally drank some thinking it was juice. At recess, to the consternation of the school's headmistress, I taught my classmates American dances such as the "bump." After school, I'd sometimes go to the nearby Jardin du Luxembourg, a magnificent park that had gardens, playgrounds, fountains, and a merry-go-round. Other days, I went to the home of my friend Genevieve whose mother would give us some baguette with a few pieces of milk chocolate inside.

My family found the fervent attachment of the French to their customs to be an endless source of amusement. When my dad asked for chocolate sauce on cassis ice cream, the waiter responded, "Impossible!" "Impossible?" my dad repeated, bemused and slightly irritated. "Non!" replied the waiter firmly, refusing to become a party to such a culinary travesty despite my father's entreaties.

On weekends, we'd visit churches and other sites throughout France. Over winter break, Andre and I went to a skiing camp in the Alps, to which we traveled along with the other campers in an overnight train filled with bunk beds. The ski runs in the Alps lasted forever and lacked clear markings, so you had to follow the skier in front of you. One day when the visibility was poor, I was following Andre, who suddenly disappeared. Moments later, I found myself tumbling down into a large crevice at the bottom of which I found Andre. We feared nobody would find us, but eventually we saw one of our ski instructors peering down at us and he helped us out.

When the school year ended, Andre and I went back to the Alps to learn mountain climbing. It was just like *The Sound of Music*: jagged mountains, verdant fields, edelweiss flowers. We learned to scale up rock faces with ropes tied around our waists. Andre was brilliant at it; I wasn't. One day, I fainted, developed a fever, and was taken to a hospital. Fearing I might have meningitis, the doctors performed a spinal tap on me, which was quite painful and traumatizing as my parents hadn't yet arrived. Fortunately, I didn't have meningitis and soon recovered.

In the fall of 1976, we returned to our home in upstate New York. I soon learned that an old town clock in Granville that hadn't worked in years was going to be torn down. It seemed a shame, so I decided to save it. I got the school to set aside the profits from the sale of food at athletic games for the repair of the clock. They were skeptical of my plan because food sales hadn't previously generated much money, but I made various changes to the operation, including selling food in the stands instead of just at the booth. Profits skyrocketed and we soon had enough money to repair the clock.

As I look back on my childhood, I feel that I greatly benefited from the variety of environments to which I was exposed. By the age of twelve, I'd attended schools in Harlem, the Upper East Side, upstate New York, and Paris. In addition, I'd spent summers in the city, upstate New York, and California, and had attended a lefty summer camp and a climbing camp in the Alps. Children benefit from change, from being forced to adapt to new situations. Our instincts are to protect them, to worry about their adjusting and making new friends. Life, however, involves curveballs and challenges, so one must learn how to adapt. Exposing kids to that early on helps them develop the skills and emotional fortitude they'll need later. That was certainly true for me.

12

CULTURE DATA

2008–2009

On August 25, 2008, six hundred new students joined Success including Dillon, whom I feared might not like it. Up to that point, he'd whiled away his days at an idyllic Jewish nursery school by dancing, doing arts and crafts, and singing Jewish songs. Not only would Success be more rigorous, he'd be the school's sole white student. After his first day, I nervously asked him how it had gone. "Good," he said. He added, however, that they hadn't sung any songs, which clearly puzzled him. "I don't think there are many Jewish kids," he said.

At Dillon's school and the two others we opened, we sought to recreate our first school's culture. This was made more difficult by our teachers' tendency to draw upon their experiences at other schools. For example, they didn't tell anyone that they were often waiting outside because the custodian at one of our schools was chronically late. When I got wind of this, I fixed it immediately, but the teachers had tolerated this state of affairs for several weeks because they'd assumed from their prior experiences that this was just one of those things they had to put up with because schools are dysfunctional. Their own low expectations had created a self-fulfilling prophecy, and such low expectations can spread like cancer. A teacher doesn't report that the lights in her classroom are out, another teacher sees the lights are out and assumes that's normal, so when her lights go out, she doesn't report it either, and so on.

We also had to fight some instincts that were deeply rooted in the psychology of our staff. When I visited a student named Sydney

McLeod who'd been hospitalized due to a stroke resulting from sickle cell anemia, I found that she was cheerful and optimistic despite the seriousness of her condition. Her mother, however, was worried Sydney would fall behind in her studies because we hadn't given her any schoolwork or homework, only get-well cards. I emailed Khari about this and he explained that when he'd heard of Sydney's hospitalization, "I immediately thought about giving her as much space from school as possible so she could recuperate and rest." The problem with this was that many of our students came from families that were lurching from one crisis to another: homelessness, illness, domestic violence, the death or incarceration of parents or siblings. If every crisis became a hiatus from schoolwork, these students would inevitably fall behind. Khari picked up on this quickly, acknowledging that rather than "setting the bar high," the school had let "lowered expectations get in the way of learning."

Unfortunately, some educators use poor children's circumstances as an excuse for failing to teach them. The American Federation of Teachers contends that since "a student's motivation and ability to learn are directly related to his or her health" students can only learn in "community schools" that serve children's medical needs as well as their academic ones.[10] That's nonsense. The health care that poor people receive today is far superior than the health care that the richest person in the world received over a hundred years ago when not even antibiotics were available. Thus, while health care should undoubtedly be improved for poor children, inadequate health care isn't a substantial factor in the failures of urban district schools. Even Sydney, who was hospitalized with a stroke, was able to do her homework.

Similarly, while it's often hard for poor parents to find time to help their children given the challenges they face, they can support their kids in school, if it is demanded of them. We required parents to check their children's homework and get their kids to school on time and in uniform. If they didn't, we'd call them; if that didn't work, we'd bring them in for a conference; and if they didn't show up, we'd give their child an "upstairs dismissal" at which they'd have to speak to the teacher or principal. If children were habitually late, we'd make wake-up calls. When one of our parents simply wouldn't read to her son even after she'd promised me she would, I invited her to a meeting at which there was a surprise guest: her mother, whom I'd met one day when she was picking her grandson up from school and seemed to me more responsible than the mom. When I explained the problem, Grandma was furious with her daughter and said, "This will not happen again, I'm taking charge." While I keep my "Grandma method" in a glass case labeled "Use Only in Case of Emergencies," it reflects our philosophy of not giving up. It's easy to develop a mind-set in which you say, "Well, the manual says do A, B, and C, and I've done those things so I've done my job." Our view was that you've done your job when you've succeeded. Moreover, while some parents resented the pressure at first, they felt proud when they saw the dramatic progress their children made.

One way to tell whether a leader was demanding high levels of parental investment was to stand outside right before school began: if kids were running down the block to make it on time, the message had gotten through; if they were sauntering, it hadn't. However, I needed a more systematic method of monitoring our schools, so I began tracking what I called "culture data"—latenesses, absences, uniform infractions, missing homework, incomplete reading logs, and whether our teachers were calling parents about these problems. This data helped us to manage our principals and helped them to manage their teachers. If a teacher's culture data was weak, it meant the teacher wasn't doing enough to get her students' parents on board. If an entire school's culture data was weak, the principal wasn't properly managing school culture.

Many people who talk about using data to manage schools focus on standardized tests, but they are given far too infrequently to be of much use. Monitoring a school with standardized test scores is like monitoring an airline's maintenance of its planes by seeing how often they crash.

Of course, assessing the health of our schools also meant looking at more subjective factors such as the quality of a principal's feedback to teachers on instructional practices, but culture data was like a canary in a coal mine: it was a simple, quick, objective measure of how well a principal was managing her staff. If a principal isn't getting her teachers to do simple things like call parents, she probably isn't getting her teachers to do harder things like improve their instructional practices. Thus, I watched this data like a hawk and wasn't shy about telling our leaders if they were coming up short. I was particularly troubled, for example, when I saw that third-grade attendance was slipping at Harlem 1, so I emailed the school's principal: "You need to be on them like nobody's business. Our founding families should know better."

The biggest change in our schools for the 2008–2009 school year was our new math curriculum. We did "Number Stories," our version of something called "Cognitively Guided Instruction." Take this question: "Aida had 19 apples, then she got 13 more. How many does she have now?" The old approach would be to show the kids the mechanics of how you put one number under the other, add the numbers in the first column, carry the 1, etc.... The new approach was to give the kids plastic "unifix cubes" and ask them to figure it out themselves. The first time, a child might just literally count the cubes. After a few times with similar problems, however, the child may notice that she can take the nineteen and turn it into twenty by taking one cube from the pile of thirteen cubes; so now the problem is just twenty plus twelve, which is easy. In this manner, children come to discover how addition in base 10 works and that is a very durable understanding they won't forget. This "constructivist" approach makes an even greater difference when students get to more complicated math such as adding fractions with different denominators.

Paul Fucaloro was again a great guide for us. If you told him that a student understood concept A but not concept B, he could tell you exactly how you could get a child to make the intellectual leap from one to the other. If he wanted them to subtract 82 from 143 in their head, he'd first have them count together by tens starting at 2 (i.e., 2, 12, 22, etc.). Next, he'd give them the problem and they could now see that by counting six 10s from 82 to 142 and adding one more, they could get to 143, so the answer was 61. Then he might ask them to subtract 82 from 141 and they would realize they could count six tens to get from 82 to 142 and then subtract 1 so the answer was 59.

So things were going well at the schools, but a threat was looming on the horizon. We needed new facilities for two of our schools and it soon became clear that our opponents weren't going to make getting them easy.

13

TARZAN AND JANE ARE BACK AGAIN

2008–2009

By 2008, there were more charter schools in Harlem than in any other part of the city. To give families the opportunity to learn about all of their educational alternatives—charter, district, parochial, and independent—we decided to invite all of the schools in Harlem to participate in a school choice fair and nearly one hundred agreed to do so. It took place at City College's gymnasium and what I saw when I arrived that morning moved me to tears: more than one thousand parents had come early and were waiting in line, not for a Black Friday sale or a concert or a basketball game, but for a chance to learn about educational opportunities available for their children. Ultimately, over five thousand parents came. It showed that parents in poor communities didn't send their kids to bad schools because they didn't care but because they lacked better options; and finally, we were changing that.

The unions, however, didn't like the competition and they'd identified our Achilles' heel: facilities. Not only would we need co-locations for future schools, but for our current ones as well since Harlem 2 was being kicked out of its current location to make way for a new middle school and Harlem 4 was running out of room. DOE offered to move these schools into the buildings of two schools it was closing, PS 194 and PS 241. Since many people don't like the idea of closing a public school no matter how badly it's failing its students, I feared this move would be controversial, but since we needed the space, I accepted the administration's offer.

DOE set the first hearing on its plan for March 11. Figuring that the unions would stir up opposition, I called on our parents to attend the hearing. Anticipating battles like this, I'd made it a practice to have orientation meetings for parents at which I warned them I'd need their help. While I'd fight for their children, I explained, I couldn't do it alone. "If you don't come and speak up for your kids," I said, "we will lose."

When we arrived at the hearing, we were greeted by hundreds of parents of PS 123 students wearing UFT hats. DOE was represented by Deputy Chancellor John White. One opponent commented:

> Tarzan and Jane are back again, swinging through Harlem, not with vines, but through charter schools. Tarzan is John White and Jane is Miss Moskowitz. . . . Like Tarzan and Jane, coming right through the black community . . . making everything better because the natives couldn't do it themselves.

Here's what some of our parents said:

> MICHELE CHRISTIAN: *I have a son that goes to [Success]. [At the] school he was going to before he wasn't learning nothing. I thought something was really wrong with my child. So he came to [Success]. They told me he was reading. I thought he couldn't read, I thought he couldn't write, I thought he couldn't do nothin'. They taught him.*
>
> PAMELA WATSON: *When we get there at 7:20 in the morning, our principal is there waiting to greet our children, shake their hands. Harlem Success is the first school that I knew of [where the] principal [knows] every one of the students' names. I communicate with her teachers every day. I didn't have that kind of communication with my other children's teachers.*
>
> [UNIDENTIFIED SPEAKER]: *I'm a Harlemite, born and raised. How many of us have struggled to find the right school within our community? We don't want to have to go down to the Upper West Side and the Upper East Side for our kids.*

[UNIDENTIFIED SPEAKER]: *My thirteen-year-old just got put in eighth grade and my five-year-old [at Success] is teaching her to read.*
SEAN JAMES: *I got a child in the third grade in the regular public school and my five-year-old also teaches her how to say the words out. I don't understand what's the big difference about [Success], but I know it's helping her and it's helping me too because I didn't get a chance to finish school.*
BELINDA DAVIS: *I'm not putting down public schools. I have a number of family members that teach at public schools. However, if your child is in a failing school, somebody should be mad about that. We don't have a problem with Harlem Success, we have a problem with the teachers, the principals, and the bureaucracy that failed your children.*

For me, it was poignant that these parents' wonder and joy at their children's accomplishments was often mixed with sadness upon realizing what they might have achieved with a similar education, and with pangs of guilt if they had an older child for whom they'd been unable to find a good school.

Due to a procedural snafu, DOE held another hearing and it was more of the same. "Here is Eve Markowitz," said one opponent, "dividing black parents." Another claimed Harlem's district schools were fine because they'd "produced Charlie Rangel, David Paterson, and Senator [Keith] Wright," all Harlem politicians. In fact, Wright had graduated from Fieldston, a fancy private school, Paterson from a school in the suburbs where his family had moved precisely so he didn't have to attend a district school in Harlem, and Rangel owed his success to having been turned around by the army after dropping out of high school (although he later went back to get his degree). Harlem parents weren't fooled. By March, we had many more applications from students zoned for PS 194 and PS 241 than zoned students who attended these schools.

To gin up opposition to our schools, the UFT used the community organizing group ACORN, which shamelessly exploited the fears of

African American families. Here's an interview with one of ACORN's organizers about what they were telling parents about charters:

ACORN ORGANIZER: *[Charter schools] are doing it . . . for gentrification.*
INTERVIEWER: *But the lottery—isn't that random?*
ACORN ORGANIZER: *Well, you're saying how you think a lottery should go. But they say, "No, this is our lottery system and this is how it goes. . . ." They can get white families coming into that neighborhood quicker if the Department of Education can give them schools.*
INTERVIEWER: *I didn't realize that charter schools had more white kids.*
ACORN ORGANIZER: *Well, but see, they're new, and of course they can't just come in like, you know, slam-dunk with the real intention. [O]nce they get established, then you'll see the change.*

As I write this, nearly ten years later, virtually no white children attend Success's Harlem schools, but these types of scare tactics were nonetheless quite effective.

The UFT then brought a lawsuit, along with the New York Civil Liberties Union, claiming that zoned school closures required approval from the Community Education Council. The city decided it was on thin ice legally and abandoned its plans. This was a tragic result for the children who ended up attending these schools. In 2015, the passage rate for students at these schools for both math and English was just 5 percent.

As for Success, it turned out that even though PS 241 wasn't closing, it had so few children that there was still enough room for us. This wasn't true at PS 194, however, so Harlem 2 was forced to remain at PS 123, which only had enough space for us for one more year.

Charter schools soon suffered another defeat. The law that increased our funding automatically to match the growth in district school spending was amended to deprive us of half of the increase to which we were entitled. This cost us millions of dollars and set a terrible precedent since our opponents would undoubtedly try to do the same in future years.

Not only was the UFT gaining political traction in its war on charters, the level of vitriol being directed against me personally seemed to be increasing daily. Indeed, the education blog *Chalkbeat* published a piece entitled "What Is It About Eva Moskowitz That Attracts So Many Enemies?":

> Why's there so much hate for a woman who has decided to spend her days starting schools for poor and mostly black children in Harlem? There are now many charter school operators in this city. Why focus on Moskowitz?

Now this was news I could use! The answer, the article suggested, was that I had a "style problem." "Rather than approaching the district public schools with respect," *Chalkbeat* said, I "dismiss[ed] their work as unacceptable." This frustrated me. The old adage "if you don't have anything nice to say, don't say anything at all" is certainly sage advice when one is contemplating the merits of one's mother-in-law's tuna casserole, but in the public sphere, it's important to be frank about problems so you can fix them. For doing so, however, I was increasingly called "divisive."

Then I learned the city council had scheduled hearings on a resolution to halt charter school co-locations. Klein urged me to testify: "U have to go after big," he wrote. I was afraid I'd be in for some rough treatment since the committee would be completely aligned with the UFT, but I felt it was important to make the case for co-location. On April 7, I showed up in the city hall chambers where I'd once served and waited to be called, feeling like a Christian about to be fed to the lions.

14

TAKE OFFENSE, IT'S OKAY

2009

As I waited for the city council's charter school hearings to begin, I recognized a UFT operative who was giving committee members cards that I suspected held proposed questions. This wasn't surprising. Robert Jackson, who'd succeeded me as chair of the Education Committee, had gotten the position in part by reassuring the UFT that he wasn't "in a position to evaluate" their contract. Translation: "I'm no Eva Moskowitz."

In his introductory remarks, Jackson claimed that "parents from nearby charter schools were brought in and deliberately pitted against parents of children attending . . . traditional public schools." This was obviously a reference to the PS 194 and 241 hearings. Plainly, we'd touched a raw nerve; the unions had previously held a monopoly on parent organizing.

Normally a former council member would be extended the courtesy of testifying first, but I waited in the audience for hours, using the time to gather my thoughts and steel my resolve. The committee's hostility toward charters was palpable, but I was determined to speak frankly. While I'd long been aware of the district schools' shortcomings, the daily contact I'd been having with Harlem parents had made me more aware of the human cost of these shortcomings. I'd seen firsthand the despair of parents whose children hadn't won our lottery and heard their stories about their terrible experiences in district schools. I owed it to them, I believed, to tell the truth about the failings of the district schools and about the efforts being made to prevent charter schools from offering families better choices. I testified in part as follows:

> There are currently twenty-three public charter schools in Harlem. For the first time, parents have meaningful choices. Now, however, a backlash is taking place. The system is fighting against innovation and parent choice. There is a union-political-educational complex trying to halt progress and putting the interests of adults above the interests of children.
>
> At PS 241, only 10 percent of eighth-graders passed the reading test in 2008. Council Member Jackson, you and I both live in Harlem, and we don't send our children to schools like this. No one on this committee would send their child to a school where only 10 percent of the students read on grade level. It is wrong to keep open failing schools to which we wouldn't send our own children.
>
> In the last two weeks, we've seen a new demonstration of the union-political-educational complex's power and influence. First, Albany recently raised zoned school funding while cutting charter school funding. Second, the council is considering a resolution that would make it harder to place charter schools in public school buildings. We all know where this is coming from. The union doesn't want parents having a choice between the education that its members are offering and the education offered at charter schools.
>
> From our local government, we are hearing slow down change, slow down parent choice. That is wrong, because every year we wait to offer parents the choices they deserve, is a year in which children's futures are destroyed.

After concluding my testimony, the questioning began:

CARMEN ARROYO: *Thank you, Mr. Chair. Eva, you and I didn't serve in the council together very long, I don't really have a relationship with you, so I am going to feel very free to have this conversation with you here.*

Arroyo appeared to be justifying in advance her intention to rake me over the coals. She continued:

ARROYO: *You in your testimony said, "Council Member Jackson, we both live in Harlem." For the record, do you live in Harlem?*
MOSKOWITZ: *I do.*
ARROYO: *Would you share with us a street?*
MOSKOWITZ: *I have three young children, so I would prefer not to. Are you questioning that I am telling the truth?*
ARROYO: *Yeah, I am.*
MOSKOWITZ: *That is a little offensive. I am happy to take the oath.*
ARROYO: *Take offense, it's okay.*

Arroyo spoke with unbridled contempt. I couldn't decide whether it was more offensive that she was accusing me of being dishonest or suggesting I'd be so stupid as to lie about something so easily disprovable.

ARROYO: *[Y]our arrogance about what the system should do and that charter schools are the answer is exactly what drives the conflict in a community. . . . You need to be mindful of that and hopefully come around to a different way of presenting how we should engage in this dialogue . . .*

Being lectured on civility by a woman who'd just falsely accused me of lying was a strange experience but I knew it was important to stay calm, so I answered in measured tones:

MOSKOWITZ: *If I have come off as arrogant then I apologize . . .*
ARROYO: *You have.*
MOSKOWITZ: *. . . but I would like an opportunity to explain, because I don't think it is arrogance, it is my own personal experience with District 5 schools. I went to them as a child, I had to figure out what to do as a mother, and it is my experience of the pain of wanting your kids to get a phenomenal education and being told it is that*

zoned school or nothing. It is the experience of holding a hundred and twenty-five hearings and then meeting with thousands of parents who want a good school.

ARROYO: *But, Eva, what I am referring to here is the approach that comes into a community . . . and setting up the dynamics for there to be conflict.*

MOSKOWITZ: *But how did I come in? I was raised there.*

Next up at bat was Council Member Inez Dickens from Harlem, who began by asking, "Do your children attend public school in District 5?" Here we go again, I thought. Just like Arroyo, Dickens was going to start with a personal attack. After I revealed that I sent one of my children to a district school and another to a charter school, Dickens gave up on this approach.

DICKENS: *I witnessed the adversarial situation between the parents of your charter school and the parents of PS 194 at a meeting . . .*

MOSKOWITZ: *It was emotional both for PS 194 parents [and] our parents because we are [told] we're from the outside. They're saying to themselves, since when did I get to be an outsider? I come from the Drew Hamilton Houses. I shop here. I work here.*

DICKENS: *I disagree about . . . shutting down [schools]. I think that this administration has been totally remiss in not . . . putting in the necessary resources.*

MOSKOWITZ: *At PS 194, it's $22,000 a child. PS 194 was failing when I was a kid. Parents deserve in real time something better. If you've got a kindergartner, you can't wait five years. Your kid will already have not learned to read.*

The piñata bat was then handed to Council Member Lewis Fidler:

FIDLER: *I know that you've always expressed and spoken your mind pretty clearly and forcefully and I think that you know that I do too and in not the most touchy-feely way.*

MOSKOWITZ: *I'm ready. Take off the gloves, go ahead.*
FIDLER: *I disagree with you in general. But I found particularly objectionable your demonization of the teachers in the city and their union. . . . I don't think . . . it's an issue of being afraid of competition. [Y]ou're comparing apples to bananas . . . I didn't ask why your class sizes were smaller. I just aspire to having my class sizes smaller too.*
MOSKOWITZ: *Well, you wouldn't like our schools very much because we have about twenty-seven kids [per class] in kindergarten.*

Fidler's assumption that our class sizes were smaller reflected the flaw in his approach to the hearings. Here was an opportunity to learn something from a former colleague such as why we had large class sizes or why so many parents wanted to send their children to our schools despite our large class sizes but Fidler instead just wanted to attack me.

The following day, the press coverage focused mainly on the index cards with questions, which the press called "cue cards." Randi Weingarten went on television to defend them, provoking a *New York Post* editorial titled "Hide Those Puppet Strings!"

"Handing out index cards in the middle of a hearing creates the appearance of impropriety," [Weingarten] said.

Ah, yes. The old "appearance of impropriety" dodge.

A more honest response would have been: "Handing out index cards in the middle of a hearing clues the public in on who's truly calling the City Council's shots."

Two weeks later, we held our lottery at which 5,000 applicants vied for just 450 seats. It's too bad that the members of the council's Education Committee didn't attend. They missed an opportunity to find out why so many parents were desperate to get their children into charter schools.

15

I'LL BE DAMNED IF I'M GOING TO LOSE YOU

1965–1982

While my family took refuge in upstate New York, the city we'd abandoned went through dark times. The seeds of its undoing had been planted years earlier. In 1965, the city had elected as mayor a maverick Republican congressman named John Lindsay who was handsome, athletic, and witty, a civic superhero who fought political battles by day and hobnobbed with movie stars by night. When the transit workers crippled the city on the day Lindsay took office, however, he appeased them with a new contract that precipitated an avalanche of demands by other unions. The sanitation workers got four successive pension sweeteners, the last of which allowed them to retire at 50 percent of their salary after just twenty years of service. "Goddamn," declared the negotiator for the transit workers, "if the garbage men get '20/50' so can we"—and so they did. Moreover, that 50 percent would now be based not on a worker's average earnings in recent years but on his compensation in his final year, *including overtime,* so transit workers conspired to assign overtime to colleagues on the verge of retirement.

These successes upset the order of the universe as the city's police and firemen regarded themselves as the most equal of the city's civil servants and therefore deserving of its most generous pensions. Their unions rose to the challenge. Heart disease was deemed a job-related condition allowing immediate retirement at virtually full pay, and pension investment surpluses were henceforth distributed annually rather than retained to cover shortfalls, which taxpayers would now have to cover.

The unions' tactics became ever more aggressive. Two unions, DC 37 and the Teamsters, conspired to paralyze the city one morning by strategically abandoning trucks on major highways and leaving all but two of the city's twenty-nine movable bridges in an open position to prevent traffic from crossing. This lawlessness was rewarded with an increase in prescription drug, dental, and life insurance benefits.

So great was the city's extravagance that it had to borrow heavily despite a booming economy. Then the 1973 Arab oil embargo sparked a recession, and the city's bloated municipal workforce and generous labor contracts, which had been extravagant at the best of times, became disastrous in what were now the worst of times. The city's debt spiraled out of control and was so great by 1975 that investors balked at buying the bonds the city needed to keep its deficit-spending merry-go-round running.

Mayor Abraham Beame, who inherited this mess, laid off some workers and demanded that those who remained forego an impending 6 percent pay increase that had been negotiated in better times. The unions responded with aggressive job actions: sanitation workers walked off the job; highway workers blocked traffic on the Henry Hudson Parkway; labor leaders led tens of thousands of civil servants in a raucous protest in the financial district; and police officers blocked traffic on the Brooklyn Bridge and then deflated the tires on the waiting cars.

The municipal unions considered declaring a "general strike" by the city's entire labor force. While these were common in some European nations, they were virtually unheard of in the United States, where strikes were reserved for negotiating individual labor contracts. This raised the question of whether the city would move toward a more European political model in which unions used their collective power to push the government toward a high tax/high benefits welfare state.

With the battle lines now clearly drawn, high noon took place not in New York City but in Washington, DC, where city and state officials lobbied President Ford to support a federal bailout. Ford's response was summarized in a now famous *Daily News* headline: "Ford to New York: Drop Dead." The city's plight, Ford asserted, was the result of "bad financial management" since "No city can expect to remain solvent if it allows its expenses to increase by an average of 12 percent every year, while its tax revenues are increasing by only 4 to 5 percent per year." If the city didn't want to pay its debts, said Ford, there was a simple solution: declare bankruptcy.

This option, however, was particularly problematic for New York because Wall Street's firms had underwritten the city's bonds, sold them to customers, and owned many of them either directly or through companies they managed. The financial industry would suffer a terrible blow if these obligations weren't honored and bankruptcy would imperil workers' labor contracts and pensions. After staring over the precipice, the parties reached a grand bargain: the unions would accept layoffs and forfeit wage increases; the state would raise taxes; and the federal government would provide inexpensive financing to ease the city's cash flow problem.

The city made good on its promise by laying off sixty thousand municipal workers, including a quarter of the city's teachers. While necessary, the layoffs took a severe toll not only on the workers themselves but on the city they'd served. Streets were strewn with trash, potholes multiplied, and subways came less frequently and broke down more often. The loss of police officers, a weak economy, and a heroin epidemic together fueled an explosion in crime so great that its perpetrators no longer even felt the need for concealment. In midtown Manhattan, prostitutes sold their wares openly. On the Lower East Side, addicts lined up in the street to buy heroin and cocaine from stores run out of abandoned buildings. In Harlem, dealers brazenly established open-air markets in which they shouted out the street names of their products: "Star Trek," "Jaws," and "Malcolm's Gold." Drug addicts stole car radios so frequently that rather than replace them, New Yorkers instead put "no radio" signs on their car windows. Adding insult to injury, Times Square, once heralded for its Broadway theaters, became a destination spot for seeing porn films and peep shows.

The city of *Breakfast at Tiffany's*—sophisticated, swanky, and exuberant—had become that of *Taxi Driver*—filthy, dangerous, and amoral—and its middle class began fleeing this fetid chaos in droves. Those who remained felt increasingly helpless as they witnessed their city being ravaged by crime and decay, a feeling punctuated by the police department's announcement on January 30, 1977, that a serial killer was on the loose. On March 8, he struck again. Nineteen-year-old Virginia Voskerichian had vainly held up a textbook to defend herself, but the killer's bullet had penetrated it and killed her. Two more victims fell in April, and this time the killer left a note so lurid and chilling that it might have been ripped from the pages of a crime novel:

> I am the "Son of Sam . . ." Sam loves to drink blood. "Go out and kill" commands father Sam. Behind our house some rest. Mostly young—raped and slaughtered—their blood drained—just bones now.
>
> I am the "Monster"—"Beelzebub"—the "Chubby Behemoth." I love to hunt. Prowling the streets looking for fair game—tasty meat. The women of Queens are z prettyist of all . . . I live for the hunt-my life. Blood for papa . . .
>
> Police—Let me haunt you with these words; I'll be back! I'll be back!
>
> To be interpreted as—bang, bang, bang, bang, bang—ugh!!
>
> Yours in murder
>
> *Mr. Monster*

Statistically speaking, Son of Sam's murders were a drop in the bucket in a city where five people were murdered every day. Psychologically, however, the failure of the police to capture a man who was so openly taunting them symbolized the city's inability to protect its citizens from crime.

Then, Son of Sam penned a letter to the *Daily News*:

> Hello from the gutters of N.Y.C. which are filled with dog manure, vomit, stale wine, urine and blood. Hello from the sewers of N.Y.C. which swallow up these delicacies when they are washed away by the sweeper trucks. Hello from the cracks in the sidewalks of N.Y.C. and from the ants that dwell in these cracks and feed in the dried blood of the dead that has settled into the cracks.

This lurid description of the city suggested that Son of Sam was a product of its decadence and decay, a social mutation born of its toxic muck.

Not until August 10, more than a year after he'd first struck, was Son of Sam finally caught. He greeted his captors with a question: "What took you so long?" By then, another humiliating blow had been struck. On July 13, a lightning strike and a single loose locking nut at a power station thirty-five miles north of the city led to a citywide blackout. Lawlessness and rioting ensued. Gangs of youths brazenly backed up their cars to stores and filled them with stolen electronics. On one thirty-five-block section of a major avenue in Brooklyn, vandals looted 134 stores and set 45 ablaze. In total, 550 police officers were injured, 1,616 stores were looted, 1,037 fires broke out, and 3,776 people were arrested, the largest mass arrest in city history.

This mayhem boosted the mayoral candidacy of a quirky congressman named Ed Koch who cast himself as a "liberal with sanity," which, by New York City standards, made him the law and order candidate. As New Yorkers were in the mood for some law and order, Koch won.

New York was coming dangerously close to its "last one to leave, turn off the lights" moment, but Koch appeared confident in his ability to right the ship and, like his contemporary President Reagan, was utterly unburdened by his responsibilities. When transit workers soon struck, Koch, unlike Lindsay, relished the fight. Walking across the Brooklyn Bridge in solidarity with commuters, he proclaimed, "We're not going to let these bastards bring us to our knees!" He proudly declared, "I'm not the type to get ulcers. I give them." He led with a marvelous sense of showmanship and chutzpah characterized by his trademark question, always delivered with a broad grin, "How'm I doing?"

Koch reassured New Yorkers both with his self-confident leadership and love of his city at a time when many feared big cities were becoming relics of the past whose sole inhabitants would be an urban underclass unable to afford the nirvana of suburban life. When later in life he managed to acquire a hard-to-come-by Manhattan burial spot for himself, he commented, "I don't want to leave Manhattan, even when I'm gone. This is my home." Koch used his infectious love of his city to battle the despair that had enveloped it. He was a cheerleader in dark times like Churchill or Roosevelt, albeit on a smaller scale, and managed to stem the exodus by sheer force of personality.

Koch also succeeded by acknowledging the plight of the city's middle class. He knew there was a limit to what they would endure, to the number of crimes they would suffer and the amount of taxes they would pay before they decided that enough was enough and decamped for the suburbs. For that reason, generosity toward the city's civil servants and its poorer inhabitants had to be balanced against the city's need to honor its social contract to provide the city's middle class with a measure of civility in exchange for carrying the city's tax burden. Squeeze them too much, Koch understood, and you'd kill the goose that laid the golden egg.

Like the broadcaster in the 1976 film *Network*, Koch let the city's middle class know that it was all right to say, "I'm mad as hell and I'm not going to take this anymore." He also sought to reassure them that the city was moving in the right direction by focusing on immediately visible change. One of the more irksome aspects of city life was riding subways defaced with graffiti while being assaulted with loud music emanating from boom boxes. This public flouting of authority, Koch believed, made subways feel lawless and unsafe, a perception that became reality as law-abiding citizens retreated. Koch therefore cracked down on radio use, had the subway cars repainted, and, to deter further graffiti, surrounded the train yards with two fences covered with barbed wire and German shepherds roaming between them.

Under Koch's leadership, things began looking up. My family decided we'd try to return to the city. To accomplish this, my brother, Andre, would apply to college a year early and I'd try to get into my father's alma mater, Stuyvesant. I studied endlessly for Stuyvesant's notoriously difficult test, as it would determine not only where I'd go to school but where my family would live. Fortunately, I passed and Andre got into college early, so in the summer of 1979, we moved back to the city.

When I went to Stuyvesant on September 11, 1979, for the first day of school, I was overwhelmed both by the number of students, three thousand in total, and how different they were from Granville's. The girl who had the locker next to mine sported a Mohawk, a nose ring, a huge chain around her neck, and an electric guitar on her back. I tried to strike up a conversation by asking if she played in a band. "Yeah," she said unenthusiastically. "What's its name?" I asked. "Steaming Vomit," she replied.

But in time I came to appreciate my classmates, who were bright and ambitious and came from a variety of backgrounds. I befriended Sung-Hee Suh, a girl of Korean descent who would go on to become a federal prosecutor, and Karen Klein, who had a Japanese mother and a Jewish father who was editor of the *New York Times Magazine*.

Some of Stuyvesant's teachers were superb, including Elaine Grist, a wonderful and hardworking history teacher. Many, however, were disengaged or ill-prepared. My AP physics teacher sometimes showed up to class drunk and would simply put his head on his desk and sleep, so I ended up learning physics from the Russian émigré students in this class who for some reason seemed to know it all already. This was part of a broader dysfunction at the school that manifested itself in many ways. The girls' bathroom stalls, for example, didn't have doors, so girls used the bathrooms at a hospital across the street. One year we didn't have enough basketballs, so we practiced by dribbling and shooting imaginary balls.

In my junior year, I heard about an organization helping refugees from Cambodia's civil war. I was particularly sympathetic to their plight since I remembered from my childhood how Nixon had bombed Cambodia. I decided to volunteer and was assigned a family that needed an apartment. I walked around the neighborhood in which this family wanted to live and called the numbers on For Rent signs I saw. Many of the landlords were wary of renting to refugees, but I argued that they shouldn't discriminate since we were all descendants of immigrants. I succeeded in convincing one landlord and when I returned to share the good news, every member of the family I was helping hugged me. I found it exhilarating that I'd been able to make a difference in their lives. Word soon got around that I was quite diligent and families began showing up asking in broken English for "Eva."

Then a family that I'd helped find an apartment in Sunset Park asked me to help them find a school as well. I knew from my experience at PS 36 that schools varied in quality, so I wanted to be sure I found them a good one. I wandered around Sunset Park until I found a school and went to the main office. I was ignored for a long time, and when a woman finally acknowledged my presence, she was quite unfriendly and couldn't tell me anything about the school, which seemed like a bad sign. I kept on walking and I found another school in an adjacent neighborhood where I'd looked at apartments but found they were too expensive. This time around, the person with whom I spoke was far friendlier and was able to tell me about the school, which sounded good. She also told me, however, that only children zoned for the school could go to it. I asked her what she thought of the school for which my Cambodian family was zoned and she said: "If you love your child, you wouldn't send him to that school." I visited more schools and noticed that the best ones were invariably in neighborhoods where I'd determined my Cambodian families couldn't afford to live. It struck me as terribly wrong that the quality of the education their children would receive would be determined by the apartment they could afford to rent.

In my senior year, I decided to work on Stuyvesant's yearbook and

I threw myself into the process with a particular sense of mission. I wanted the yearbook to reflect the unique spirit and character of our school and express our class's collective experience as students. When we took our faculty pictures, we encouraged them to vamp for the cameras. One posed with the bicycle he rode to work; another held up a sign with a number on it as if he were a convict; a third, who taught history, held a globe. And rather than focusing myopically on what had happened within the walls of our school, we aimed to acknowledge the impact of external events on our experience.

We laid out a section combining pictures linking these two worlds: pictures of the Three Mile Island nuclear power plant and students wearing No Nukes T-shirts; of the historic King Tut and Picasso shows and of students painting; of the launching of the space shuttle and of students studying science; of John Lennon, who'd been murdered, and of students playing music; of Prince Charles and Lady Di marrying and of students kissing. Other sections had collections of wonderful color photos of students captured in a variety of activities and were accompanied by essays about our time together and quotes from authors ranging from Dr. Seuss to Sartre to the Doors.

When the yearbook was finally complete, the vendor brought me a copy and we paged through it together. He'd initially resented all the additional work we were making for him, but he now told me that it had been the best professional experience of his life and that he'd never look at yearbooks in the same way. I learned that while people may initially dislike being pushed hard, they may feel differently when they see the results of their labors. I also discovered the satisfaction that comes from working with talented people. I wasn't a particularly good writer or photographer, nor did I know anything about layout. My sole talent was getting other people to use theirs: inspiring them, organizing them, holding them accountable. Together, we'd shown that if you put your heart into it, you could take something that might otherwise be pedestrian and unimportant and make it great and meaningful.

Working on Stuyvesant's yearbook, I became close friends with a classmate named Sarah Nazimova who, in addition to being a very talented artist, was incredibly bright and interesting. We'd talk for hours on end about our thoughts on everything: education, psychotherapy, politics, friendship, love. Through her, I got to know a classmate named Eric Grannis who one day invited me to a picnic in Central Park. He seemed interesting and witty so I accepted his invitation. We got together a few times over the ensuing months and then he invited me over to his place for dinner on a weekend when his parents were away. Since Eric had platonic friendships with several girls, I wasn't sure his interest was romantic, but I suspected it was when he served me a sophisticated Italian meal he'd cooked (chicken with porcini mushrooms, risotto, and a mocha gelato) at a candlelit table. A good night kiss removed all doubt.

A week or so later, we were walking in a park near school and began talking about education, a topic that Eric had thought about a great deal as his father was a professor of education. Eric thought students at failing public schools should be given tuition vouchers to attend private schools. I'd never heard of this idea, which was quite obscure then, and it deeply troubled me. Education, I said, was far too important to be handed over to the private sector. In that case, asked Eric, was I against food stamps? No, I said. Well, he replied, tuition vouchers were just food stamps for education. If I trusted private industry to make food, why not schools? And conversely, if I thought the government was more competent than the private sector, shouldn't the government produce food for poor people rather than just give them food stamps? Was I also against all government aid for students attending private colleges or did my distrust of private education only extend to K–12 education and, if so, why? I found Eric's ideas disquieting but I was attracted to his intellectual independence and quick mind.

While Eric was very smart, he wasn't terribly focused. He'd been an indifferent student and seemed to be taking a rather lackadaisical approach to getting into college. Whereas I'd studied endlessly for the SATs, Eric had taken them cold. My head therefore told me that I should be wary of him. As for my heart, I'll let my diary tell the story:

2/2: "Think that I'm beginning to get emotionally involved with Eric. I enjoy talking to him and he's a very special person."

3/12: "God, I'm falling in love with Eric. All I talk about with Sung-Hee is Eric, about what an amazing person he is."

3/13: "I'm afraid I'm going to mess things up. It's scary to love someone."

I think what attracted me to Eric was his love of learning. He wrote short stories and plays, took jazz lessons, learned how to cook gourmet meals, and read fiction copiously; he did all this not so he could get into a prestigious college or impress people or make a lot of money but simply because he enjoyed it. He also seemed to be interested in me for my mind. He'd taken months to get to know me before he'd even kissed me on the cheek and for my birthday had given me a collection of short stories by Flannery O'Connor, one of his favorite authors. While I don't think Eric gave a moment's thought to how to make me fall in love with him, he couldn't have formulated a better plan if he'd tried.

Eric was intrigued when I told him about bike trips I'd led as a camp counselor so we decided to take such a trip together the summer after our graduation. We biked from Montreal, Canada, to Buffalo, New York, sleeping in a tent and cooking our own meals. From Buffalo, we flew to Florida to visit Eric's aunt Franny. I'd assumed she'd be a short old woman who played mah-jongg, but when we arrived, out walked a glamorous middle-aged woman in a white bikini. It turned out she was a counterculture poet who'd written a racy coming-of-age novel titled *American Made* and then married a millionaire businessman with whom she'd lived in a sixty-room mansion. Now divorced, she split her time between Manhattan's West Village and the Florida Keys where, she informed us, pot was plentiful and of high quality because it washed ashore when dumped by smugglers running from the Coast Guard.

Eric's family, it turned out, was from a quite different background than mine. His mother, Alexandra, was the daughter of an Irish poet who had eloped with a debutante whose ancestors were Huguenots who'd come to America in the eighteenth century. Eric's father, Joe, a bookish man who'd initially studied for the ministry at Harvard before going into education, was from a protestant family in Milwaukee. Yet, while our families had come from strikingly different backgrounds, they had similar values. Our parents were all educators—Eric's mother, Alexandra, was a learning disabilities specialist—who valued the life of the mind and enjoyed nature. These similar values helped bring Eric and me together.

I worried, however, about what would happen to our relationship now that we were heading off to different colleges: the University of Pennsylvania in my case, Columbia College in Eric's. After we parted, I wrote Eric a long letter reminiscing about our relationship and sharing my worries about our future: "I'm scared of being hurt—you finding another woman that you love more. I'll be damned if I'm going to lose you. I feel very satisfied but completely unsatiated. I want more. I love you. Eva."

16

EVEN OLYMPIC ATHLETES CRY

2009

Many people have observed that standardized tests don't measure real learning, just superficial test-taking skills and rote learning, and that our country's obsession with testing is doing profound damage to our educational system. If that's what you believe, I'm afraid my views may disappoint you. I believe well-designed standardized tests measure real learning and understanding.

Consider the following math questions:

How many positive two-digit numbers are evenly divisible by 4?

The measures of the angles of a triangle are in the ratio 1:2:3. What is the measure of the largest angle?

If x is an integer, what is the greatest possible value of the expression $1-x^2$?

A boy has as many sisters as brothers but his sister has twice as many brothers as sisters. How many boys and girls are there in the family?

These are all good math questions because they require creative thinking and an understanding of conceptual math. They are questions that an effective teacher might teach in class or put on a test but they also can and do appear on standardized tests. Requiring a student to select among five answer choices and fill in a bubble doesn't make these problems any less challenging or worthwhile.

Standardized tests can also measure a student's ability to read carefully and thoughtfully, to understand complex passages, and to analyze what they have read. Test makers have even figured out how to measure many of the important skills associated with good writing. I'm not talking merely about rudimentary subject-verb agreement, but also about subtler skills such as recognizing vague references to antecedent nouns, identifying a proper vocabulary word by making subtle distinctions in meaning, and knowing how to make sentences more precise and less awkward.

Ironically, it's actually teacher-created tests that more often focus on rote learning (e.g., "Who was America's first president?") or simple procedural skills ("What is 125÷5?"). Standardized tests are better designed because a team of professionals can put far more effort into creating a test than an individual teacher. The College Board's accomplishments with the Common Core–aligned SAT are particularly impressive.

It's a myth that students can do well on a standardized test just by learning superficial "test-taking skills." Sure, skills like learning how to pace yourself affect your performance, but you still need to know your stuff. Take the math questions above. No test-taking skill or trick allows you to answer them correctly without mastering mathematical content. SAT prep courses may help you raise your scores by teaching you new vocabulary words or problem-solving techniques, but that's still real learning. When my oldest son studied for the SATs, it had a noticeably positive impact on his vocabulary.

But forget my opinion. American universities, which are the envy of the world, make standardized test scores one of the greatest single factors in admissions for both their undergraduate and graduate programs. No law requires that they do so; they choose to because, in their experience, these tests are a strong measure of a student's academic accomplishments. Surely if standardized tests can measure whether a student can handle Harvard College or Yale Law School, they can measure a fourth-grader's math and reading skills.

Many teachers don't like standardized tests and don't want their students to be anxious about them, so they tell their students that the tests aren't valid and they shouldn't worry about them. Such a message, however, inevitably diminishes students' motivation to do well on tests that are very important to their future. This isn't a problem for the children of affluent parents who, however much they may rail against standardized tests, will make sure that their own children do well on them, but it is a huge problem for poor kids who rely upon their schools to prepare them.

If you're a teacher who doesn't like standardized tests, by all means advocate for your position. Petition Congress. Write to universities. Publish op-ed pieces. But please don't tell your students that tests don't matter, because you're just selling them a bill of goods. Your duty is to prepare your students for the world as it exists, not as you would like it to be. Moreover, while telling kids tests don't matter might make them feel good in the short term, they'll feel differently when they don't get into the college they want to attend or don't have the career to which they aspire.

Our students would be taking standardized tests in the spring of 2009 for the first time, and I was determined to give them as good a chance of doing well on these tests as students from affluent families. I was particularly concerned for Sydney, the girl who had been hospitalized for sickle cell anemia. Unfortunately, she'd gotten pneumonia and spent additional weeks at the hospital and then had a relapse as soon as she was released. She hadn't been able to return to school until October 20 and even then was weak.

I initially left test preparation to my teachers, but when I saw what they were doing, I became increasingly concerned. They were giving students dubious strategies and advice such as not to change an answer because one's first choice is usually right. I believe the best way to prepare students for a reading comprehension test is to teach them how to comprehend what they are reading. Isn't that just regular school? Yes, pretty much. The difference is really about intensity of effort, about using the prospect of competition as an opportunity to dig deeper and try harder.

Kids are naturally impulsive, so they read a passage as if they were taking a Rorschach test: something in it catches their interest or reminds them of something and so they think the passage is about that. Take the nursery rhyme "Jack Sprat": "Jack Sprat could eat no fat. His wife could eat no lean. And so between them both, you see, they licked the platter clean." A child might say that this rhyme's main idea is that different people like to eat different things since that's how the rhyme begins. That interpretation, however, leaves out the poem's entire second half. A better statement of the main idea is that people who are different can make a good team because they complement each other. Or take the nursery rhyme "Humpty Dumpty." A child may say it's about the need to be careful with things that can break, but that doesn't account for the last two lines: "All the king's horses and all the king's men couldn't put Humpty together again." A better description of the main idea is that some things can't always be fixed when they break.

Thus, our job was to get students to move past their initial reactions and instead to consider the whole text. Distinguishing a text's main idea from subsidiary arguments often requires understanding the logical structure of a text. For example, the main idea of the Declaration of Independence isn't that all men are created equal, although that's a catchy line, but that Britain's violation of Americans' inalienable political rights justified revolution.

One of the most popular teaching techniques nowadays is to ask students to make what are called "text-to-self connections," such as "the boy in this story gets mad at his sister like I do sometimes." While children benefit from this approach when they are quite young, as they mature they need to be able to focus on what the author means, not only on their own reaction to the text.

Many critics of standardized tests claim test preparation is a waste of time. I agree that bad test preparation is a waste of time, but when it is done right, it can be quite intellectually valuable. In fact, I have found over the years that our students actually learn more when they do test preparation than at other times of the year because both they and our teachers are so focused on mastery.

Our students took the state reading test in January and we then turned our attention to preparing them for the math test. Again, this didn't mean learning tricks or superficial strategies. Rather, it involved making sure the kids understood the necessary mathematical concepts and learned how to be careful in their work. Just as in reading, the kids were naturally impulsive and careless. Suppose for example that a child multiplies 3.5 by 3.6 and gets 1.26. When the child sees that 1.26 is one of the multiple-choice answers, he concludes he must have gotten the calculation right when, in reality, he's made a decimal point error the test maker has predicted. We therefore taught children to double-check their work. That, however, didn't mean just doing the same thing again but rather using another method. For example, in the problem above, the student would estimate the answer by figuring out that it must be somewhere between 9 (3 times 3) and 16 (4 times 4). Estimating is both a useful skill and, in this context, requires that the student understand the mathematical principle that the product of two larger numbers will always be greater than the product of two smaller numbers.

The other piece of the puzzle was teaching children to be careful and precise. Paul wrote to me in February that he was deeply troubled by the "general tolerance/acceptance [of] sloppiness/carelessness in all of our schools, a lack of insistence on accuracy, precision, and attention to detail." If a student was sloppy when he copied numbers from a word problem, Paul would make him practice that skill until he learned to be more careful. Learning to be meticulous in your work is a critical life skill whether you're taking a standardized test, performing heart surgery, or planning a space mission.

We also taught kids to try harder because often the problem wasn't a lack of understanding but a lack of effort. For example, one of our teachers emailed me about a student who had "bombed practice today, but while we are sitting here one-on-one he is slamming it." We could also tell a child wasn't trying when he spent the final minutes of the exam staring off into space at the end of a test rather than double-checking his answers.

Our job was to get the kids to care about their academic work just as much as they cared about other things in their life such as being popular or playing basketball well. To do this, we'd praise students for improving their scores and confront them if they were being lazy. Did this sometimes make them feel bad? Of course! That's normal. If you want to go through life never feeling bad, don't aspire to accomplish anything, root for a sports team, or fall in love. Caring inevitably means feeling bad at times. In fact, some of our students cared so much they'd cry when they didn't do well. We didn't encourage that but neither did we see it as a sign that something was horribly wrong. Watch eight-year-olds lose a championship basketball game: they cry. Even Olympic athletes cry, but nobody goes around saying, "Oh, those Olympic athletes shouldn't care so much."

Good teaching is like good parenting: love is necessary but so too is disapprobation. That doesn't mean humiliation. A teacher should never say "You're stupid" or "You'll never amount to anything." In fact, they should say the opposite: "I'm really disappointed in you because I know you have the ability to do so much better than this." Teachers must never give in to unhelpful emotions such as anger or frustration. A teacher who loses control loses trust because students know that out-of-control adults are prone to hurting them either physically or emotionally. While it's perfectly normal for teachers to become angry and frustrated, they need to let off steam with their friends or a therapist, not students. Khari, who was particularly insightful about emotional relationships with students, noticed that some teachers were unnecessarily escalating conflicts to satisfy their own emotional needs:

> It is not a good idea to deliver bad news to scholars in the moment. Instead of telling a scholar prone to tantrums, "I am calling home!"

just call home. Instead of telling a scholar, "You now have upstairs dismissal," just take the scholar to upstairs dismissal. Do not tell a scholar, "I am calling Mr. Shabazz." Just call me.

Scholars do not have to know what you are going to do in the heat of the moment, and sticking it to them in front of their classmates will almost always lead to a battle you can't win.

The key to using emotions is for the teacher to express ones that show she cares for her students, wants them to do well, and believes they can succeed. As long as that happens, teachers can and should express both positive emotions such as joy and love, and negative ones such as disapproval, sternness, and disappointment. This can be particularly effective with students from troubled homes because they often need an adult in their lives who acts like one: who is patient, who doesn't let his emotions get the better of him, who sets clear rules and expectations, and who metes out love and disapprobation in a reasonable and predictable manner. Paul was masterful at this. He didn't shout or get angry but was nonetheless very direct. When students didn't do well, Paul let them know in a way that made them feel they'd let him down.

About a month or so before the state tests, Paul told every teacher at Harlem 1 to send him their most challenging students, the ones who were misbehaving or making little progress academically, so he could teach them. The teachers thought Paul was nuts, that it would be chaos if you put all of these students together in a single class, but he soon had them eating out of his hand and they made dramatic academic progress.

As the testing date came closer, we started having test prep on Saturdays. I made sure every student attended and personally intervened when necessary. I wrote an email to my staff about my conversation with one parent: "Spoke to mom. She apologized. Disorganized, etc. She says he is in school now. Told her he must come both Saturdays and I will personally come to her apartment and drag her and him out of bed if he's not there."

On May 7, we got our results. Ninety-five percent of our students had passed the English Language Arts (ELA) test and 100 percent of our students had passed the math test, compared to 76 percent and 93 percent statewide and 56 percent and 82 percent in District 5, Central Harlem. On the math test, 71 percent of our students had gotten 4s, the highest score, compared to 27 percent statewide and 15 percent in District 5. Miraculously, even Sydney, who had missed more than a month of school and continued to suffer from serious health issues, had not only passed both the ELA and math tests but had gotten a 4 in ELA.

I was very proud of both the teachers and the students and was enormously relieved. First, and most important, it meant that, at least on some level, we were succeeding in teaching children. Standardized tests don't measure everything, but they measure something. You can't pass a reading test if you can't read. In addition, I believed that doing well on these tests would strengthen our students emotionally. Many of them suffered from things over which they had no control: homelessness, absent fathers, abused mothers, troubled siblings. They had now learned that they could succeed at something to which they put their minds, that they did have some ability to control their lives. Lastly, strong test scores meant we could get approval to open more schools so we could serve more students.

To do that, however, we'd also need more space, and in the course of looking for that space, we'd noticed something curious: while district school enrollment in Harlem was declining, it was doing so much more slowly than charter school enrollment was increasing, which made no sense given that we were pulling from the same pool of students. It wasn't until we pored over the data that we got to the bottom of this mystery.

17

FROEBEL'S GIFT

2009

District schools, we figured out, were propping up their declining enrollment by admitting more students to pre-K. As a result, the worse a school was, the more pre-K spots it added. I complained to DOE that they were allowing these failing schools "to harm younger and younger kids" but the head of DOE's pre-K responded that, au contraire, these programs must be great because "enrollment is high." In fact, enrollment was high because charters weren't allowed to offer pre-K, something that pained me deeply since I could see how little our students had learned in the district pre-K programs they'd attended before enrolling at Success. In pre-K, children should learn basic social skills such as sharing, taking turns, cleaning up after themselves, and expressing their needs with words; to count and to recognize shapes, colors, and sizes; artistic skills such as painting, cutting with scissors, and building with blocks; musical skills such as singing and playing a rhythmic instrument; and movement skills such as dancing, jumping, and hopping. Our kids were learning very little of this in the district pre-K programs they were attending. Most of these programs were instead just wildly expensive babysitting.

But one day, Eric told me he might have a solution to this problem. He'd noticed that state law said schools could educate children "between the ages of four and six" in kindergarten although they could also "fix a higher minimum age." Why, Eric wondered, would the law contemplate that such a broad age range be taught in a single grade? He found the answer in the history of kindergarten.

By 1837, Friedrich Froebel had taught in a boarding school, written books and pamphlets on education, run an orphanage, and fought in the Prussian army that had defeated Napoleon at Waterloo, but his most important work was still ahead of him. That year he founded a "Play and Activity Institute" in which young children sang, danced, gardened, and played with educational toys Froebel had invented, which came to be called "Fröbel-Gaben" or Froebel Gifts. In time, Froebel dubbed his invention "kindergarten"—a garden for children—and, with the help of a devoted proponent and patron, the Baroness Bertha Marie von Marenholtz-Bülow, his ideas spread far and wide, eventually reaching America where, in 1856, Margarethe Meyer Schurz, one of Froebel's disciples, opened a kindergarten in Waterton, Wisconsin, that was conducted in German.

Fascinating, I said to Eric, but what does this have to do with the price of tea in China? He explained that the 1947 New York law authorizing kindergarten gave school districts flexibility to "fix a . . . minimum age" for kindergarten because educators then didn't think of it as just an additional grade tacked on before "first" grade, but rather, like Froebel, as a multiyear program akin to what we today call nursery school. Indeed, New York's education commissioner had ruled in 1974 that school districts could run two levels of kindergarten: a regular kindergarten to prepare kids for first grade and a "developmental kindergarten" program for kids who weren't yet ready for the regular kindergarten.

Since the law allowed charter schools to offer "kindergarten," this meant we too could offer "developmental kindergarten" (or "DK" as we dubbed it). I loved this idea not only because it would allow us to start educating children as soon as they turned four, but it would also allow us the flexibility to assign kids, regardless of age, to either developmental or regular kindergarten, which would be helpful since children mature at different rates. This discovery had the potential of being a huge deal for the charter sector.

We asked both of our authorizers to amend our charter so we could offer DK, including in our submissions a supporting letter from Professor David Steiner, who was the dean of the Hunter School of Education and sat on Success's board. Ed Cox, chair of SUNY's charter schools subcommittee, liked our idea but wanted to be sure we were on firm ground legally, so he consulted the head of the New York City Charter School Center, James Merriman, a leading authority on charter school law, who concluded we were right about DK. So too did Carl Hayden, a distinguished attorney who was the head of the SUNY Board of Regents and had previously served as commissioner of the State Education Department (SED). On January 16, 2009, SUNY approved our request to offer DK at our three SUNY-authorized schools.

But strangely, we didn't hear a peep from SED for months on end. They simply refused to act. In April 2009, I asked SED's newly appointed chancellor, Merryl Tisch, for help. She responded, "Eva, your work speaks for itself. The only thing we should be doing is encouraging you[,] not getting in your way." Music to my ears! Moreover, we soon learned SED's new commissioner would be David Steiner—yes, the very same David Steiner who'd written a letter of support for DK!—and that the regents were going to officially appoint him on July 27, the very same date on which our application would be considered.

However, I soon began hearing rumors that opposition to DK was building. The teachers' unions didn't want charters to be able to compete with the district schools by offering pre-K and they had a lot of influence with SED. On July 24, Associate Commissioner Shelia Evans-Tranumn told me SED wasn't going to approve DK because it was illegal. I began calling SED's regents to ask for their support, but while they liked the DK concept, they felt their hands were tied since they were being told it was illegal. I complained to Merryl Tisch that while SED was telling its regents that DK was illegal, it had "never given me a legal opinion nor has any person from SED ever pointed to a single specific provision of the law that supports this position." It was Kafkaesque: SED claimed that by offering DK I'd be breaking the law but wouldn't tell me which one.

Then I heard from Carl Hayden, head of SUNY's board, that SED was threatening to revoke our charters if SUNY let us run DK, and from James Merriman that other charter schools were told that votes on their requests were going to be delayed unless they got me to withdraw my DK application. Why go to all this trouble to pressure me to withdraw my application when SED could just vote it down? Because they were afraid that if I went public, it would be obvious that politics was at work since SED would be voting down a program that SED's brand-new commissioner had previously supported.

Ordinarily, I wouldn't have hesitated to take on SED but I didn't want to undermine David or Merryl, who were both education reformers who could do a lot of good for us in the long run even if they weren't willing to fight for our DK program. Moreover, even if SED wouldn't approve our DK program, SUNY had, so we'd at least be able to offer it at our three SUNY-authorized schools. I wrote Tisch that I wouldn't mount a public fight if the regents rejected our DK program: "It deeply pains me because it is truly wrong for our kids but I will sit quietly and leave." While I could justify my decision on strategic grounds, I knew I was selling out the kids who'd been selected for Harlem 1's DK program, and for the commission of this sin, I sentenced myself to witnessing in person SED's execution of our DK program at their meeting in Buffalo.

After the meeting, SED finally put its objections to DK in writing. DK, it claimed, was illegal because it would "prepare students for . . . regular kindergarten[,] not . . . first grade." This position was flatly contrary to New York law and SED's 1974 decision under that law that schools could offer "developmental kindergarten" for students who needed an additional year of kindergarten before first grade.

While SED vetoed SUNY's approval of this program at our SUNY-authorized schools, SUNY had the power to override SED. To make sure SUNY did, I sought the opinion of yet another lawyer, Joseph Wayland, who was a demigod among educators because he'd brought a famous lawsuit establishing that the funding for

New York's public schools was constitutionally inadequate. Wayland concluded that SED had "fail[ed] to provide any legal authority supporting SED's position" and had "ignored the applicable law described above, as well as the Charter School Act's policy favoring innovation." And mind you, he gave us this opinion pro bono, not as a hired gun.

At this point, a virtual who's who of lawyers had determined that DK was legal: Wayland, James Merriman (head of the Charter School Center), Ralph Rossi (counsel to SUNY's charter schools institute), Ed Cox (head of SUNY's authorizing committee), Carl Hayden (head of SUNY's board and a former SED chancellor), Joel Klein (New York City's chancellor and the former head of the Antitrust Division of the Department of Justice), and of course, my favorite lawyer of all, Eric, who'd come up with the idea in the first place.

With Wayland's memo in hand, SUNY hung tough and overrode SED's veto, but SED had one more card to play. It sent a letter to Joel Klein warning him that "no State Aid should be allocated to . . . developmental kindergarten." This meant the city would have to pay for our entire program out of its own pocket. I prevailed on Klein to do so for the upcoming school year but I doubted the city would do so forever.

We opened up DK at Harlem 2 and it was quite successful but, just as I'd feared, Klein soon emailed me that he couldn't keep on paying for it "unless I know I can get reimbursement" from the state. I asked Wayland if he could bring a lawsuit to make the state pay for DK. "It has involved so much risk and angst and political capital," I explained, that I'd "have scuttled [it] long ago" if it weren't "so great for our kids and such a compelling innovation." Wayland replied, however, that a lawsuit could take years and it was always possible we'd lose, which meant we'd be stuck with the costs of the program. I reluctantly decided the risk was too great and we'd need to end the program. "Breaks my heart," I wrote Harlem 2's principal, "because you have done such a great job." It was one of the most painful things

I've had to do. The result was wrong in every way: educationally, legally, and morally.

Coincidentally, around this time Shelia Evans-Tranumn, one of DK's assassins, retired from SED to become executive director of the UFT Charter School. I was curious to see how she'd do now that she was actually responsible for educating children herself and not just telling others how to do so.

18

I'M SUCH A LUCKY SO AND SO

1982–1994

After college began, Eric broke up with me. Since his own parents had gotten divorced, he explained, he wanted to decide carefully whom to marry. He felt he was too young and inexperienced to make that decision now, and that if he wasn't willing to make that commitment, it wouldn't make sense to let our relationship go on indefinitely.

I was heartbroken. Eric was everything to me and it hurt me deeply that I wasn't everything to him. I felt bereft and alone. I also found it difficult to relate to my classmates who seemed to have a lot of money and to view college primarily as a means of getting more of it. I was probably being unfair, but I had romantic ideas about college and the pursuit of knowledge. Yet, adding insult to injury, I found myself struggling in my coursework. My papers, I was told, were verbose, poorly structured, and replete with rookie mistakes like the use of the passive voice. My teachers at Stuyvesant had never warned me my writing skills were weak. To the contrary, I'd always gotten excellent grades. This seemed terribly unfair. I was quite a diligent student and would have worked to improve my writing if I'd been warned it was lacking. Instead, they'd sent me off to college unprepared. Luckily, Penn had a writing clinic and I visited it religiously.

A bright spot for me was Penn's great faculty, particularly those who taught history, my favorite subject, including Professors Lee Benson, Bruce Kuklick, Michael Katz, and Drew Faust, now president of Harvard. They were engaging, brilliant, and passionate about their work. I loved studying and did so incessantly. At the library, I'd lose track of time and had to scurry out when they started turning off the lights. I particularly loved reading primary sources and piecing together history from archival materials. For a paper on leftist student politics in the 1930s, I reviewed a huge trove of student Communist newspapers in the New York Public Library archives. One day an old man asked me why I was reading them. It turned out he'd been a Communist student in the 1930s, so I interviewed him and he then introduced me to some other former Communists who were also happy to be interviewed. I was in historian heaven!

I was disappointed by the political apathy of my fellow students at Penn, so I created an organization called the Political Participation Project to encourage students to become involved in politics. I also became active in campus groups protesting the issues of the day: apartheid, the CIA's involvement in Nicaragua, and excessive sentences for drug crimes.

I'd stayed in touch with Eric both because I valued him as a friend and thought we might eventually get back together. Rather than date other women, Eric had spent all of his time studying, becoming almost obsessively intellectual. Then, during our sophomore year, Eric suggested we get back together. The distance may have helped him. I think Eric had become so worried about hurting me that it had overwhelmed his positive feelings about our relationship. Maybe the fact that I'd managed to move on had reassured him that I wasn't as fragile as he feared. Whatever the reason, I still loved Eric, so I let him undump me and we hatched a plan to spend our junior year together at the University of Edinburgh in Scotland. I enjoyed the university, at which we studied English and history, but the best part was traveling on the Continent, which we did every chance we got. Over winter break, we spent three weeks visiting northern Europe, seeing van Eyck's astonishing altarpiece at Ghent, Amsterdam's incredible Van Gogh Museum, and the beautiful medieval town of Bruges where, walking the empty streets on Christmas morning after a light snow had fallen, it felt like we'd traveled back in time.

During our spring break, we spent a month biking in Portugal,

Spain, and France, and then another two months bicycling through Greece and Italy that summer. We slept most nights in a tent and Eric would use a camp stove to cook a simple but delicious meal, often pasta with garlic, olive oil, and Parmesan or perhaps a tomato sauce, and I'd make a salad. This, along with a bottle of local wine and an appetite spurred by miles of biking, made for a feast. We'd often ask people if we could camp on their property, which frequently prompted an invitation to join them for dinner or drinks. This gave us the opportunity to meet people that most tourists wouldn't encounter, such as farmers. One old man in the south of Italy showed us a textbook he'd used in elementary school that, to my amazement, had sections on every subject: math, history, geography, art, literature, science. It was all in one book because that was all they could afford.

In Greece, we were so moved by seeing the Parthenon that we visited it again the following day. In Italy, we loved the Coliseum in Rome, the cathedral at Orvieto, the leaning tower of Pisa, and bicycling through the Tuscan and Umbrian countryside with its enchanting hilltop towns. In Siena, we met up with a family friend who took us to see the blessing of a horse that would represent his wife's neighborhood in the Palio, a race on the cobblestones of the central square that had been held annually for nearly four hundred years. The horse we saw blessed won, so we were able to join in the festivities, a bacchanalian outdoor feast that went on until the wee hours of the morning.

We soaked it all in just as my parents had two decades earlier. Even now, I can remember large portions of these trips virtually day by day. It's hard to convey just what a magical experience it was. Imagine spending months bicycling through foreign countries, eating delicious meals, seeing beautiful landscapes, visiting historic sites, exploring ancient towns, and doing all of this in the prime of your youth and with the love of your life.

When we returned to the United States, I decided to apply to graduate school given my love for studying history. I chose Johns Hopkins, which had an outstanding program and gave me a generous financial aid package. Eric taught science at a public school in East Harlem during my first year, but got a teaching job in Baltimore the following year. That summer, the Iran-Contra hearings took place and we couldn't tear ourselves away from them. They featured a collection of characters straight from central casting: the pipe-smoking Admiral John Poindexter; Ollie North with his boyish American grin and ramrod-straight posture; Fawn Hall, the beautiful and loyal aide who'd helped him destroy documents, in one case by secreting them out under her dress; North's fiercely combative counsel, Washington legal powerhouse Brendan Sullivan who, when a senator complained that he was making too many objections, famously shot back, "I am not a potted plant!"; and Arthur Liman, a brilliant and quirky New York lawyer whose manner was once said to be "Walter Matthau doing a Perry Mason impersonation." When North impishly observed that he thought he'd destroyed all copies of a document Liman was grilling him about, Liman replied, "Colonel, my eyesight has suffered from reading what you left behind." At the time, watching these hearings felt like an indulgence, but doing so came in handy later when I came to chair a committee.

Eric was interested in politics so the following year he got a job working for Harlem congressman Charlie Rangel and we moved to DC, where I continued working on my dissertation. We lived on Capitol Hill. On Saturday nights, I'd keep Eric company while he cooked, drinking wine and listening to a wonderful public radio show called *Hot Jazz Saturday Night*. Afterward, we'd stroll down Constitution Avenue to the Capitol and look out over the mall.

I found writing my dissertation challenging. Seeing me struggle, Eric suggested one day that I should go into politics since I was good with people. I dismissed the idea. I could no more imagine myself as a politician than as an astronaut.

As for Eric, although he had a lot of interests—politics, writing, investing—he didn't seem to be pursuing any concrete career plans and I feared that I wouldn't be able to count on him to be much of a breadwinner. I asked my mother whether this should dissuade me from marrying him. It shouldn't, she said, if I truly loved him, which I did. In 1989, Eric proposed, although that's a charitable description; there was no ring or bended knee or romantic spot, just a suggestion Eric made at a diner as casually as if he were proposing we go to a movie. But knowing Eric was clueless in these matters, I took his lack of calculation as a mark of sincerity rather than a sign of indifference.

We were wed on July 2, 1989, on the lawn of my parents' house in upstate New York under a chuppah my father had made from branches cut from trees on our property. Our wedding processional was a recording by Duke Ellington, whose music we'd come to love from listening to *Hot Jazz Saturday Night,* and the ceremony ended with the upbeat Ellington song "I'm Such a Lucky So and So." It was a beautiful summer day, and while I suppose that, just as all infants are cute, all weddings are beautiful, I do think there was something especially magical about ours.

And, in keeping with my family's tradition of parsimony, it was cheap! Soup to nuts, it cost $600, although that was more, I suppose, than my parents' half a wedding cake and day trip to Sausalito. We'd bought our clothes from thrift stores, Eric cooked for our ninety guests, Eric's aunt Franny baked a cake (which had two little bicyclists at the top!), and a friend of ours made tablecloths from lace and red fabric (though I admit I think my father took matters too far when, having been sent to the store for crushed ice and apparently seeing no need for such an extravagance, he returned with cubes and a hammer).

My parents were happy with Eric, as he had all of the qualities of a good Jewish boy other than actually being Jewish. Grandma Frances told me, however, that she disapproved of my marrying a goy. I was hurt but not resentful. I didn't take lightly the fact that I was breaking a chain of Jewish marriages that had probably lasted for millennia; I just loved Eric.

For our honeymoon, we of course took a bicycle trip, this time in the south of France, which had beautiful countryside, delicious food, and a great network of small roads. We biked from Nice to the beautiful city of Avignon and back in a large circle, enjoying Provence's Roman ruins, quaint towns, and wonderful cuisine.

That fall, I continued working on my dissertation and managed to get a one-year stint as an assistant professor at the University of Virginia. Since I was just twenty-five, my youthful appearance made for some awkward moments. When I sat down at the round seminar table on the first day of class, a student asked me if I'd "heard anything about Moskowitz."

One day, Eric suddenly announced he'd decided to apply to law school. Predictably, he had decided to do so not because he'd figured out what he'd do with a law degree, but rather because his interest in the law had been piqued by some articles on constitutional law that had come across his desk at work. Eric got into Columbia Law School so in August of 1990, we returned to New York City.

In our absence, New York had continued the upward trajectory it had begun in the Koch years. Wall Street was booming and its wealth trickled down to innumerable other businesses, from law and accounting firms to car dealerships and restaurants. The trickle, however, dried up before it reached the bottom: unemployment in poor neighborhoods was high and homelessness was rampant. Moreover, a crack epidemic had pushed the murder rate to record levels. There was a growing sense that while Koch had been the man for the time, that time had passed. The city chose a Harlem politician by the name of David Dinkins to replace him.

I didn't know what to expect from Eric in law school since his level of studiousness had fluctuated dramatically. It turned out that Eric's instinct that he'd enjoy law school was correct and he thrived as never before. At the end of the year, he finished among the top five students in his class and he got a summer position with Cleary Gottlieb, arguably the country's best international law firm. In his final year, Eric won the law school's moot court competition and got a prestigious position as a clerk to a federal district court judge.

I was glad to see Eric fulfilling his potential but I felt I wasn't. I'd earned my PhD in record time but had been unable to find a permanent teaching position. I'd found a one-year stint at a college in New Jersey quite dispiriting since the students had no enthusiasm for history. I was trying to turn my dissertation into a book but was finding it hard. Eric encouraged me to look into other careers. I made a documentary about how ordinary women experienced the changes brought about by the women's lib movement. Years later, it was picked up by a video distributor that sold it to many universities, but nobody was interested in it at the time so my prospects as a documentarian looked poor. I sunk into depression. I was so used to having a sense of direction and purpose that I felt lost without them. All of my self-confidence and self-esteem drained out of me. I began seeing a psychotherapist twice a week. It didn't help. I had no idea of how to get out of this hole or if I ever would.

In 1993, I managed to get a job at Vanderbilt University. While it was just a one-year position, Vanderbilt was so prestigious that it could be a springboard for a permanent position elsewhere. I became more optimistic about an academic career. While it was difficult to be away from Eric, who was doing his clerkship in New York City, I enjoyed teaching at Vanderbilt as the students were enthusiastic and bright and my colleagues were friendly and capable.

Then I was informed that a permanent position had opened up and was encouraged to apply. This should have been a no-brainer but I was reluctant to do so as I feared I'd never truly feel at home in Nashville. My hesitation raised a more profound question: whether I was really committed to academic life and the sacrifices it entailed. Most academics were willing to live anywhere for a chance to pursue the life of the mind. I wasn't so sure. While I enjoyed learning about history, I still found writing a struggle. Moreover, an article I'd written had been rejected for publication for what I believed were political reasons. I feared I'd have to conform to academia's politically correct culture to advance.

I struggled with whether to withdraw my name from consideration, as my diary entries from this time show:

February 6: I want to escape but will I be happy once I do? Will I find something else? I don't have the same kind of self-doubt I had a year ago, but will [it] come back? Withdraw[ing] feels somehow suicidal. Could I find a niche for myself? How do you make a difference in society?

February 7: Going back and forth feeling tormented by this decision.

February 22: Spent weekend with Eric. We decided I will be coming to NYC to write my book.

February 24: I'm turning thirty and I don't have a job. But other people have taken a while to find themselves.

February 26: Sometimes I feel so inadequate in this profession. All these intellectuals expect me to know all this stuff that I don't know about. I long to feel competent and confident. Is there another profession that I would feel better about? Or is this my psychological destiny?

February 28: Looking at my letter [withdrawing my name] made me anxious. It feels like such a gamble. Heading into the unknown is so scary.

19

AN EVA MORATORIUM

2009

Unlike many unions, which exercise power with a bludgeon, the UFT was actually quite sophisticated. Much like a good film director who can use subtle cinematic techniques like camera angles to manipulate your reactions, the UFT could control events without most people even realizing it was pulling the strings. I knew something was afoot, however, when on June 2, 2009, Harlem 2's principal emailed me that protestors were outside the school chanting "Whose school? Our school!" and "HSA, go away!" Since this was occurring right outside our classrooms and was upsetting to our young students, our principal "ask[ed] them politely to move," but they "shouted [him] down."

Daily News columnist Juan González covered the protest, reporting a parent's claim that "half the current fifth-graders at PS 123 have been reassigned . . . to make room for . . . the Moskowitz charter school." This was completely false. González hadn't in fact spoken to a single parent who claimed they'd been forced to take their child out of PS 123. Rather, he'd interviewed a parent who had speculated about why other children had left. In reality, fifth-graders who graduated from PS 123's elementary school weren't required to attend its middle school and many had simply chosen not to as there were better alternatives.

González struck again on July 2 with an article about our taking over rooms from PS 123:

> No one was expecting the moving men when they arrived Thursday morning at PS 123 in Harlem.
>
> Not Principal Beverly Lewis, nor any of her staff, nor any of the school's parent leaders.
>
> "These strangers suddenly appeared, went up to the third floor, removed the cylinder locks from a bunch of classroom doors and started moving out all the furniture and computers, and piling everything up in the gym," said one teacher . . .
>
> The moving men claimed they had orders . . . to make way for an expansion of the Harlem Success Academy.

In a subsequent story, one of González's colleagues repeated González's claim that "Eva Moskowitz's Success Academy had movers break locks off doors."

What really happened on July 2? I'll let my email exchange with González tell the story. It began with an email from González at 2:19 p.m.:

> I'm told that workmen hired by Harlem Success came to the school today, drilled off the locks on a half dozen classrooms and began removing furniture . . . without notifying the principal or anyone else in the building.

I replied:

> You have not been given the correct facts. There is a space allocation agreement which allocated the rooms in question to Success. On May 18, Tom Taratko, director of space planning at DOE, wrote to Dr. Lewis and me: "This allocation will take effect on July 1st." Similarly, on June 30, Timothy George of DOE wrote to the custodian, copying Dr. Lewis, "Dear All: This agreement is in place and shall take effect tomorrow July 1st."
>
> Indeed, Dr. Lewis herself recently acknowledged the July 1 date in an email.

> We did not drill off the locks. In fact, the custodian opened up the locks for us with a key.

González responded:

> I personally saw holes in the doors where the locks would normally go. . . . If the custodian had opened the doors with a key, the locks would presumably still be there.

I responded:

> The custodian removed the locks. This is DOE's standard operating procedure when a new tenant moves in.

DOE does this for security reasons; it doesn't want the old school to have the keys for classrooms it's no longer using.

Everything I told González was reflected in the written record or could be confirmed by DOE but he nonetheless went ahead with his story, claiming that our move was unauthorized and we'd drilled out the locks. He didn't even publish our denials that we'd drilled out the locks or mention the emails we'd gotten authorizing our move. The only reference to our position in the article was a statement that "Moskowitz denies impropriety," which made it sound like I wasn't denying having drilled out the locks without authorization but just thought it was fine and dandy to do so. Even when we got written statements from DOE that they had removed the locks, not us, González refused to correct his story.

In dealing with dozens of journalists who'd written more than a hundred articles about me, I'd never encountered one who was as willing as González to print outright lies to advance his ideological agenda. Moreover, this wasn't just about my feelings being hurt (although they certainly were, since González's article portrayed me as lawless). The bad press could be used as ammunition to limit colocations or cause DOE to delay our getting access to these rooms, which we needed so we could renovate them before our teachers arrived in early August.

González's articles also engaged in ad hominem attacks on me. He wrote that my "critics, who include educators, parents, the teachers' union and Harlem political leaders, say she is a relentless self-promoter." Writing that I'm viewed as a bad person by all sorts of unnamed good people (educators, parents . . . probably nuns and animal lovers too) isn't journalism, it's character assassination.

I wanted to fight back to stop this endless string of lies, but I was advised that suing or objecting publicly wouldn't do any good and would likely just make the misleading coverage worse, so I dropped the matter. It had become clear to me by this point that some in the media had an accountability problem. Usually, people in our society are held accountable for their wrongful conduct either by the press or by the legal system, but neither works well for the press because journalists are reluctant to criticize their colleagues and libel lawsuits are generally a fool's errand. I've therefore tried to follow Benjamin Franklin's advice not to pick fights with those who buy ink by the barrel. Since many people follow that advice, however, the press often gets away with reckless and biased reporting, which is problematic given their importance to our society. As George Orwell observed, "He who controls the past controls the future." If people believe that charter schools have been harmful in the past, then charters won't be allowed to expand in the future. For many years, I refrained from publicly criticizing the press. I've decided, however, that the press is too important for me to remain silent.

Five days after González's article, another protest was held outside PS 123. One person held a sign that said "Eva, go home! Pigs at the trough!" with a picture of me as a pig. The education blog *Chalkbeat* noted that ACORN organizers were at this rally, which was the UFT go-to front group. In the yearlong period ending two weeks after this protest, the UFT paid ACORN and its successor $325,616.

On July 10, ACORN staged yet another protest at PS 123 which was attended by my former colleague Tony Avella, who repeated the false claim that we were pushing out PS 123 students, and my former opponent Scott Stringer, who claimed that we were "strangers to the community" despite the fact that most of our parents, including me, lived in Harlem.

That day, the union's endgame was revealed: state legislation was introduced to limit the mayor's power to co-locate charters. The unions understood that if parent demand was the fuel for charter school growth, facilities were the oxygen because they allowed charters to bypass the lengthy capital-intensive process of building new school facilities. The UFT had chosen its moment well because a critical law that gave Bloomberg more control over the school system than his predecessors was set to expire. The assembly, which was controlled by the UFT, could hold the renewal of this law hostage unless legislation was passed to limit co-locations.

The UFT also wanted to restrict Bloomberg's ability to close failing schools. When a school closed, its teachers remained on DOE's payroll but couldn't actually teach until another school chose them. The costs of these unwanted teachers had reached $200 million annually. Pressure was growing to stop paying them, which would effectively mean the end of tenure for teachers at failing schools.

The UFT had figured out how to kill two birds with one stone. It got its allies in the assembly to propose a law requiring that any significant change in a school building's utilization—i.e., a co-location or a school closure—be subject to a complicated approval process involving multiple hearings and the publication of an "Educational Impact Statement." The UFT had borrowed this idea from the environmental laws that require the publication of "Environmental Impact Statements" for public work projects, which lawyers had used to strangle projects with red tape. In one famous case, a highway and park project in New York City called Westway was delayed for more than a decade, at which point the city had just given up on it. I feared the UFT would use this law to similar effect since, as the UFT itself eagerly observed, the law would require DOE to disclose "a long list of specific details" such as the "effect . . . on personnel needs, the costs of instruction, administration, transportation, and other support services" every time it wanted to change a school's utilization.[11] DOE would inevitably make mistakes in preparing these detailed statements, which the UFT could then exploit to bring lawsuits. The whole thing was perfectly constructed to gum up the works, but on its surface—and this was the UFT's brilliance—it had that fresh minty taste of transparency and good government that could be sold to the media as reform. On July 16, this proposed legislation, which also contained other troubling provisions that would weaken mayoral control, was voted out of committee but the following day a dozen Democrats joined Republicans in voting it down.

Harlem's state senator, Bill Perkins, was pushing another bill, dubbed the "Eva Moratorium," that would limit the number of charter schools that a single network could open. Perkins claimed my co-locations were causing conflict: "You have two parents . . . and all of a sudden here comes Eva Moskowitz to take over their school, unauthorized" and the "parents are fighting their neighbors." Perkins was claiming my actions were unauthorized based on González's false reporting. As for parents fighting, that happened only because ACORN and the UFT would manipulate district school parents by telling them we were secretly planning to gentrify their neighborhood or going to increase the class size in their schools.

None of the anti-charter proposals had enough support to pass the Senate, but Bloomberg couldn't afford a stalemate because mayoral control was expiring. He therefore concluded that he had no choice but to accept the proposed limitations on co-locations and school closures, in exchange for renewing mayoral control. The resulting compromise was enacted on August 11.

This episode troubled me because it showed just how adept the UFT had become at getting legislation passed. It didn't just hire lobbyists and make campaign contributions. It paid ACORN to stage protests, then called politically aligned journalists like González to write about the protests, then had politicians it supported hold hearings scripted with cue cards, then had other politicians introduce legislation that the UFT had drafted before the show had even begun but which would now be justified as necessary in light of the protests, stories, and hearings the UFT had manufactured. It was frighteningly effective.

To this point, I'd stayed out of Albany politics, preferring to focus on running my schools, but I decided I could no longer do so. It had been a terrible year for us politically. Moreover, we were soon going to reach the cap on the number of charter schools allowed under state law. If we couldn't get it lifted, New York's charter school movement would be stuck in neutral. Things looked grim. However, events were playing out at the national level that could have a profound impact on our fight in Albany.

20

THE AUDACITY OF HUBRIS

2008–2009

There had long been signs that Barack Obama held views on education considered heretical in Democratic circles. On the topic of tuition vouchers, he'd said we should "see if the experiment works, and if it does, whatever my preconception, you do what's best for the kids." While doing "what's best for the kids" might seem to be an unassailable proposition, vouchers were considered so abhorrent in Democratic circles that failing to express horror and revulsion at their mere mention was considered suspect. Not surprisingly, Obama had become more circumspect about his views on school choice as a presidential candidate, but his principal education advisor, Jon Schnur, was firmly in the education reform camp.

As an Obama victory looked increasingly likely, Democrats for Education Reform (DFER), the organization founded by John Petry, began preparing a memo to Obama's transition committee that would suggest people to appoint and policies to advance. In mid-October, DFER's executive director, Joe Williams, whom I knew from the time he'd been a *Daily News* reporter, emailed me a draft of DFER's memo. Joe observed:

> This is going to be one of the greatest inside baseball battles in edu-history. . . . [W]e've shared . . . our push list with people like Wendy Kopp, Ted Mitchell, Joel Klein, as well as key staffers from George Miller and Ted Kennedy's staffs. . . . A lot of where this goes will depend on Jon Schnur, and all of this is designed

> to give him cover within the campaign. Joel Klein also will play a key role in rallying Broad, Gates, and prominent Dem donors behind whomever our coalition eventually gets behind for ed secretary.

The day after Obama won, Joe got troubling news: Stanford professor Linda Darling-Hammond, an ideological ally of teachers' unions, would be on Obama's transition team. Her appointment particularly incensed Whitney Tilson since Darling-Hammond was perhaps best known for a study criticizing Teach For America, which Whitney had helped found. His immediate instinct was to pen a blistering blog post since, well, that was always his immediate instinct. Fearing, however, that this would simply chase the Obama administration into the arms of the unions, Joe made Whitney hold his fire until Joe could find out from Schnur what Darling-Hammond's appointment meant. Schnur told Joe not to panic because transition roles were being given out as consolation prizes to people who weren't going to get administration positions. Armed with this information, Joe got Whitney to stand down and made his final revisions to his memo, which he sent to Obama's transition team on November 7. For Secretary of Education, Joe recommended either Chicago school superintendent Arne Duncan or Teach For America founder Wendy Kopp. The latter suggestion was a Joe Williams knuckleball: he knew Kopp wouldn't get the nod but hoped she'd draw Darling-Hammond's fire away from Duncan.

DFER also recommended people for dozens of other positions, arguing that "A 'team of rivals' approach . . . in which the administration seeks to placate/nullify all of the various education interests by giving everyone a seat at the table, will do very little to advance any sort of agenda which could possibly be considered 'change.'" Translation: we don't want a seat at the table, we want *all* the seats. When DFER's memo became public, one critic, riffing off the title of Obama's book, *The Audacity of Hope*, pronounced Joe's memo "the audacity of hubris." Maybe—but it worked. Not only was Duncan appointed, so too were many of the other people DFER recommended.

The Obama administration's main focus, however, was the financial meltdown caused by the mortgage-backed securities debacle. Lehman Brothers had gone bankrupt, AIG and Citibank had been saved from the same fate only by a $100 billion government bailout, the stock market had suffered its worst week since the Great Depression, and Fannie Mae had dodged insolvency through a government takeover. The chairman of the Federal Reserve warned Congress that unless it passed Bush's $700 billion emergency bailout immediately, "we may not have an economy on Monday."

Congress complied, but far more government spending was needed to pull the US economy out of its nosedive. This plight put the normal Washington budget dynamic on its head. The green eyeshades who normally tried to rein in spending, but were now looking for ways to increase it, asked Schnur if he could find a way to take, say, $100 billion off of their hands. Indeed he could! He came up with the idea of creating a competitive grants program that would reward states for adopting educational reforms such as expanding access to charter schools and shutting down failing schools.

The challenge was that few Democratic congressmen supported educational reform. One who did, however, was House Education and Labor Committee chairman George Miller, who was also an early Obama supporter and friendly with Schnur, whom he'd gotten to know when Schnur was associate director for education policy for Bill Clinton. Miller liked Schnur's proposal and helped sell it to David Obey, the chair of the Appropriations Committee. Obey wasn't wild about the competitive grants program but he did like the sound of $100 billion in education spending, so they agreed on a compromise: $5 billion would go to Schnur's grants competition and the remaining $95 billion would be distributed to district schools. Normally Schnur's proposal would have been heavily scrutinized and probably killed by the teachers' unions, but there was so much pressure to pass Obama's $800 billion economic stimulus package that Schnur's program, dubbed Race to the Top, was swept forward like a twig on a tsunami.

This was a huge win for DFER, and while Joe Williams had led DFER brilliantly, one could be forgiven for thinking a higher power supported education reform given the lucky breaks DFER had gotten: Obama's early contact with DFER, his improbable election, Duncan's friendship with Obama, Schnur's role as an advisor to Obama, Miller's ideological leanings and relationship with Schnur, and, most of all, the financial crisis, which, albeit awful for the country, had unleashed $100 billion in education spending.

For my part, I certainly hoped this program would help with the politics in New York because we were about to reach the charter school cap and the teachers' unions were becoming increasingly aggressive in their efforts to stop charters from expanding.

21

WITH WHOSE PERMISSION?

2009–2010

"There is nothing more difficult to plan, more doubtful of success, nor more dangerous to manage, than a new system," observed Machiavelli, since "the initiator has the enmity of all who would profit by the preservation of the old institution and merely lukewarm defenders in those who gain by the new one." This applied to the charter movement because legions of teachers and union members vigorously defended the district schools, which educated 97 percent of public school students in New York, and while many people supported charter schools in the abstract, they weren't going to march in the street for our cause. That we'd even made it this far was sheer luck. Then governor George Pataki had gotten the legislature to pass the first charter school law by promising them a pay raise for doing so. Then, by the time we reached the cap on the number of charter schools allowed to open under the original law, Bloomberg, an avid supporter of charters, was at the height of his power.

But luck doesn't last forever. One day, a less sympathetic mayor or governor would be elected, and we'd be in deep trouble if we hadn't gotten big enough to stand on our own two feet. This fact was not lost on the UFT. They knew that the bigger we got, the harder it would be to stop us, and by 2009, they'd decided to draw a line in the sand by opposing a second cap lift. Moreover, our strongest ally, Governor Eliot Spitzer, had resigned when his dalliances with a high-end call girl came to light. While his successor, David Paterson, also supported charters, he lacked Spitzer's political skills and instinct for the jugular.

Seeing an opening, the unions were using every tool at their disposal to oppose a cap lift: mobilizing their members; calling on legislators whose campaigns they'd funded; using ACORN and other front groups to stage parent rallies; planting stories with ideologically aligned reporters; and using their long-standing relationships with legislative staff members, many of whom had actually once worked for the UFT. Taking on the UFT would require challenging them on every front: fighting a ground war in the streets of New York City and an air war in the halls of Albany; soliciting news stories praising the accomplishments of charters and editorials decrying our opponents' schemes to do us in; asking legislators who were sitting on the fence to take our side and pressing those who already had to take a firm stand.

That our opponents weren't letting up for a moment became clear on September 9, when they staged yet another protest outside PS 123, shouting slogans such as "No justice, no peace," "Whose school? Our school!", "Eva Moskowitz must go," and "Eva Moskowitz is a poverty pimp." They wouldn't even give a safe corridor to our students who were forced to "dodge[] and weave[] between demonstrators to get into the building."[12] The *Daily News* reporter who covered this event from her desk claimed it was a parent rally. It wasn't, as you can see for yourself on YouTube.[13] In addition to teachers holding up signs such as "teachers don't 4get," the protesters' signs and T-shirts identified them as members of various left-wing groups such as the "Class Struggle Education Workers," the "PPP," the "No To Mayoral Control Coalition," and the "Internationalist Group," which "fights for international socialist revolution, the conquest of power by the working class, led by its Leninist party, championing the cause of all the oppressed." We later learned that this protest had grown out of a meeting that State Senator Bill Perkins had hosted.

Our opponents continued to mine co-location drama for anti-charter articles. González wrote about an AstroTurf soccer field we'd built on a playground we shared with a co-located school. Although replacing asphalt with a more forgiving AstroTurf field that both schools could use would seem to be in everyone's interest, González claimed that a complaint two parents had made about the field was proof there were "bitter space wars between charter schools and regular public schools."

Our opponents also claimed we were depriving district schools of needed space. Here is a typical example:

> A teacher from PS 241 painted a dire picture Tuesday of what happened to her school after Harlem Success Academy opened in her building in September 2009. "They have squeezed us and suffocated us almost into oblivion."[14]

Similarly, state senator Tom Duane, a perennial opponent, claimed that "[Success] aggressively annexed essential classroom and cluster room space" at PS 241.[15]

Instead of investigating the accuracy of these claims, reporters usually just reported what each side said. Printing demonstrably untrue claims was apparently fair play, as long as they reported our denial. Now perhaps you're thinking, "Well, Eva, maybe it's not as black-and-white as you claim." Okay. Let's go through the data and you can decide for yourself.

Every year, New York City publishes a document called the Blue Book that reports on the occupancy of the city's public school buildings. It shows that many of the city's most popular district schools function at more than 100 percent of their target capacity including PS 303 (181 percent), PS 51 (174 percent), PS 228 (173 percent), and PS 242 (165 percent). How is that possible? Do they have classes in the hallways? No, because the Blue Book's target capacity calculation isn't like an occupancy limit for a movie theater or an elevator. Capacity isn't how many students a school can hold if every room is packed to the gills. Rather, it takes into account the need to have extra rooms for science labs, art, counseling, and special education. A school at 100 percent capacity is like a couple with one child living in a two-bedroom apartment: it's just the right size.

Now let's look at the buildings Success occupies. I'm going to fast-forward to 2014 because we'd opened up more schools by then so there's more data. According to the Blue Book, PS 241, the school I was supposedly suffocating, was in a building that could comfortably serve 1,118 students but its enrollment was 99 students. No, that's not a typo. Absent a co-located school, it would have been at 9 percent capacity and virtually every student could have had his own private classroom.

"Okay, Eva," you say, "but how much of that space did PS 241 get to use after you took all their rooms?" Great question! The space allocated to PS 241, the Blue Book says, was sufficient for 192 students. Thus, their facilities would have been comfortable for double their actual enrollment.

"Sure, Eva," you say, "you've cherry-picked one good example. But I wasn't born yesterday. What about your other schools?" Okay, Woodstein, here is a chart concerning every single co-location we had in Harlem in the 2013–2014 school year:

NAME OF SCHOOL	CAPACITY OF BUILDING	DISTRICT SCHOOL ENROLLMENT	EXCESS SEATS AVAILABLE	UTILIZATION RATE OF DISTRICT SCHOOL(S) BEFORE CO-LOCATION	UTILIZATION RATE OF DISTRICT SCHOOL(S) AFTER CO-LOCATION
WADLEIGH	1,512	960	552	63%	81%
PS 241	1,118	99	1,019	9%	52%
MS 149	551	80	471	15%	49%
MS 207	660	339	321	51%	89%
MS 208	699	41	658	6%	68%
MS 099	1,345	205	1,140	15%	15%
MS 101	819	46	773	6%	66%
MS 030	1,156	432	724	37%	75%
PS 123	934	602	332	64%	89%
MS 175	706	391	315	55%	96%
AVERAGE	**950**	**320**	**631**	**32%**	**68%**

This chart shows that, absent co-location, these buildings would be at a third of their capacity on average, and that even with co-location they were only two-thirds full. In fact, buildings that have charter schools in them are actually *less* crowded on average than schools that don't: 76 percent versus 104 percent.[16] And these are *facts*, not opinions: you can look for yourself at the Blue Book, which was put out by a later mayor who wasn't a big fan of charters, with the input of "principals . . . , the United Federation of Teachers, community groups, and elected officials."

Now, perhaps you're thinking, "Okay, Eva, but even if you don't cause *over*crowding, you must admit that you make these schools *more* crowded than they'd otherwise be." Actually, no, because we don't grow our students in test tubes, we just serve the children who would be attending neighborhood public schools anyway. That was why the enrollment of students in Harlem district schools was shrinking. In fact, while opponents claimed we caused an increase in class size, the opposite was generally true. Thanks in part to yours truly, PS 241's average class size fell to—wait for it—eleven students. Eleven!

But despite these facts, reporters continued to print stories that undermined not only our efforts to get the charter cap lifted but also to get co-locations for two new schools we were opening and for Harlem 2, which was running out of space at PS 123.

On November 26, DOE told us that it planned to co-locate Harlem 2 with Kappa II, a school DOE was phasing out, but I feared the city might abandon this plan if the opposition became too great. These fears increased on January 1, 2010, when Bill de Blasio, a former city council colleague who'd been elected public advocate, announced that his office would engage in "community organizing" to help parents keep Kappa II open.[17] The late *Village Voice* columnist Wayne Barrett noted that there were two groups of parents whose fates were at issue—those of Kappa II and those of Success Academy 2—and yet de Blasio was only offering to help one of them, the group whose interests just happened to align with those of the UFT, which had given de Blasio $12,000 in campaign contributions.[18]

Back in Albany, the city's effort to raise the cap was being led by a twenty-eight-year-old wunderkind named Micah Lasher who had graduated from Stuyvesant where he'd run the student newspaper, which the faculty had shut down after Micah wrote a column saying that teachers shouldn't be appointed to Stuyvesant based on seniority. The paper had been allowed to reopen after a story ran in the *Times* noting that Micah had accused the school's administration of "blackmail," and presented it with a petition signed by half the student body. Coincidentally, Micah had been witness to the beginning of my political career as a fourteen-year-old volunteer on my first campaign and had presided over its end as campaign manager for the candidate who'd bested me in my final campaign. But one can't afford to hold grudges in politics; I was just happy Micah was on my side this time around.

We had one thing going for us: under Obama's Race to the Top grant competition, New York could win a $700 million grant if it adopted education reform measures such as raising the cap on charters. To remove the cheese without springing the trap, our opponents proposed legislation that would raise the cap on charters but then hobble us in other ways that would slow our growth, such as giving a district school's parents the right to veto a co-location in their building. This sounded pro-democracy, but it would create a rigged vote since neighborhood residents who might want a charter school option but didn't happen to send their child to the district school in question wouldn't be allowed to vote. Our opponents also proposed ending SUNY's power to grant charters.

Governor Paterson called a special legislative session for January 18, the day before the federal grant application was due, to pass procharter legislation. Perkins vigorously opposed this legislation, claiming that there was an "oversaturation of charter schools," despite the fact that there were thousands of applicants on charter school waiting lists. He also maintained that charters were "separate, and inherently unequal," implicitly likening them to segregated schools. While this was a stretch given that we selected our students by lottery rather than race, Perkins knew that parents whose children were stuck in failing schools found it terribly painful to see other children get advantages denied to their own, whatever the reason. Perkins's perverse solution to this situation, however, was not to expand charters, so that more children could get a quality education, but to restrict them so fewer could.

Many people oppose charter school expansion because they believe it will ultimately hurt children and, while I fervently disagree, I appreciate their concerns. Perkins's opposition, however, stemmed from a different place as another charter school leader explained in a *Times* article:

> Geoffrey Canada, who . . . has a cordial relationship with the senator, said that much of Mr. Perkins's ire seemed to stem from the fact that many people running Harlem's charter schools are not from Harlem.
>
> "When people say they are going to scale up . . . in Harlem, I think the question for some political leaders is, 'With whose permission?'" Mr. Canada said.

Since Perkins had commented on another occasion, "Eva Moskowitz . . . acts as if she doesn't need any authorization to do things," I assume Perkins considered me one of these people who are "not from Harlem" although I lived and worked in Harlem, sent two of my children to schools there, and had actually attended a District 5 school myself. Canada, on the other hand, had graduated from a school in the suburbs of Long Island and now lived there and yet Perkins didn't have a problem with his running schools in Harlem. Canada, however, was African American and that seemed to be what Perkins really cared about.

Perkins's opposition to school choice was particularly curious given that he'd attended the exclusive private school Collegiate, and Brown University. Despite receiving this first-class education, he never hesitated to take the low road, accusing Bloomberg, for example, of "treating us like we're some people on his plantation." Perkins epitomized Henry Adams's dim view that "politics, as a practice . . . [is] the systematic organization of hatreds."

The legislature failed to pass any pro-charter legislation but New York submitted its federal grant application anyway. I hoped the application would be rejected since that would put pressure on the legislature to pass pro-charter legislation to strengthen a second round application.

To keep the pressure on, we asked our parents to go to Albany on February 3 to meet with their representatives. "My son read a book last week called *Caps for Sale*," responded one parent. "He is five. I was not able to read at that level till I was seven. Please save a seat for me on the 6:30 a.m. bus." After attending a rally with hundreds of other charter school parents, our parents met with several legislators. One legislator who refused to meet was Perkins. Parents were so fed up with his repeatedly dodging their requests that they tracked him down and confronted him in the hallway. If Perkins had instructed his chief of staff to prevent their meeting with him, said one parent sarcastically, "You need to triple her pay, 'cause she does her job."

While the air war took place in Albany, the ground war in the city raged on. At a February 22 hearing on Harlem 2's proposed colocation, opponents claimed the city was closing Kappa II to make room for Success. In fact, the city was closing Kappa II because only 9 percent of its students were reading at grade level. I, however, was a convenient villain. "It's a shame," said one opponent, "that people like Eva Moskowitz are coming in to divide our black and Latino community." At another co-location hearing, ACORN operative Jonathan Westin was present to make sure their anti-charter message was repeated. "[A]fter the school," said one speaker, "they're coming after your apartments." While "chairing the Education Committee," said another, I'd been secretly making "plans to take over Harlem and our public schools." Council Member Inez Dickens accused us of engaging in "hostile slash and burn corporate takeover tactics that would be envied by Genghis Khan."

Using scenes like this, the UFT maintained that co-locations were "pitting parents against parents." It was brilliant. The UFT would rile parents up by telling them we were going to increase class size at their schools or were responsible for plans to close them, and would then blame us for the resulting strife.

We fought back by emphasizing how much demand there was for charter schools, particularly ours, which had received 7,000 applications for 1,100 spots. In some Harlem zip codes, 63 percent of age-eligible children had applied to Success. For a school we were opening in the Bronx, we had twenty-eight applicants per seat.

On February 26, Governor Paterson announced that he wouldn't be running for reelection that fall due in part to a brewing scandal involving a romance with a longtime aide. Paterson's lame-duck status made him even weaker and our opponents seized on this by trying to push through legislation on March 12 that, while raising the cap, contained poison pills that would leave charters worse off overall. Fortunately, it didn't pass but the vote was too close for comfort. The teachers union soon announced it would be targeting five state senators who'd opposed the legislation including a courageous freshman senator named Craig Johnson whose potential challenger was offered a $200,000 war chest.

Our only hope was that the federal government would reject New York's grant application and I was increasingly fearful it wouldn't. Hardly any states had passed pro-charter legislation so the bar for a winning application might be set quite low. Secretary Duncan, however, said that wouldn't happen. "This is not a race to the middle," he declared, "this is a race to the top," and he made good on his rhetoric on March 29 by rejecting all but two of the applications that had been submitted including New York's.

The next round of applications was due on June 1 and we ratcheted up the pressure in various ways including soliciting favorable newspaper editorials that said that the legislature would be responsible for losing $700 million in federal money if they didn't pass pro-charter legislation. The elephant in the room was Attorney General Andrew Cuomo, who'd announced he was running for governor and was virtually certain to win. Both sides had been lobbying him to take a stand and on May 23, he did. He declared that the cap should be lifted, that SED shouldn't get a monopoly on charter authorizing, that "New York must be the leader [in] education reform," and that he opposed any restrictions on co-locations that would be "a poison pill that prevents opening new charter schools." We were elated by this full-throated support. This wasn't the type of mealy-mouthed pronouncements that politicians typically make on controversial issues.

With the June 1 deadline rapidly approaching, the parties were in virtually continuous negotiations. My name was coming up so much, one of the city's negotiators told me, that the entire dispute could probably be settled by promising never to give me another co-location. Sheldon Silver, the assembly speaker, was refusing to lift the cap without anti-co-location provisions, and on May 27, the city walked out of the negotiations. It looked like there really wouldn't be a deal after all. Then Silver blinked. He called the city back to negotiations and gave up on his strongest anti-co-location demands. At 10:30 p.m. that night, the parties announced they'd reached a deal. It required DOE to prepare a "Building Utilization Plan" that would lay out in detail exactly how the schools would share space and would require DOE to spend matching funds on a district school when a co-located charter school fixed up its facilities. I wasn't wild about either of these requirements. The Building Utilization Plan was just one more chance for DOE to make mistakes that the UFT could exploit in a lawsuit. As for the matching funds, I was all for the city fixing up its facilities but I worried the end result might be to prohibit us in the future from using private funds to fix up ours because the city didn't want to match our expenditures. These provisions, however, were far less damaging than the poison pills our opponents had proposed, and we'd gotten the cap lifted, we could keep on growing. That wouldn't have happened without Obama's Race to the Top program. His support was absolutely crucial to the growth of charter schools at this critical time in our history.

That fall, state senator Craig Johnson's support for charter schools cost him his seat. While he only served one term, he used it to provide critical support on two major educational issues, mayoral control and the cap lift. In my mind, it's better to be a one-term legislator who votes his conscience than a veteran legislator who betrays it.

22

DOING SOMETHING WORTH WRITING

1994–1999

In the summer of 1994, I returned to New York City, which had elected mayor Rudolph Giuliani in my absence. Such is the nature of politics that Giuliani probably wouldn't have been elected but for a car accident that occurred on August 19, 1991. The final car of a motorcade carrying the leader of the Chabad-Lubavitch Hasidic religious sect accidentally struck and killed seven-year-old Gavin Cato, the son of Guyanese immigrants. Hundreds of mainly African American protesters began shouting, "Jews! Jews! Jews!" and three days of rioting ensued in which an Israeli flag was burned, 27 cars were destroyed, 152 police officers were injured, 225 robberies and burglaries were committed, and, most troubling of all, a young man named Yankel Rosenbaum was fatally stabbed. Many Jews believed Mayor Dinkins had gone easy on the protesters for political reasons. While that probably wasn't true, his restrained personality didn't serve him well in situations like these. New Yorkers preferred mayors like Koch and La Guardia whose brash and larger-than-life personalities suggested they could tame the city.

In hindsight, however, Dinkins served the city well. When he took office, the murder rate had been growing steadily for three decades. Dinkins increased the size of the police force and appointed a police commissioner, Ray Kelly, who adopted a community policing approach that contributed to an astonishing reduction in crime in the years that followed. Dinkins also cleaned up Times Square, decreased homelessness to its lowest level in twenty years, ran a largely scandal-free administration, and conducted himself both in office and afterward with considerable dignity and goodwill.

When I returned to the city, I didn't know what direction my career would take but I did manage to finish my academic book *In Therapy We Trust: America's Obsession with Self-Fulfillment,* a history of psychological thinking in American culture. It began with Phineas Quimby, an early nineteenth-century spiritual teacher and mesmerist who invented a "talking cure," and went through the TV talk shows and twelve-step programs of the 1980s and 1990s. I felt adrift and it frightened me.

For as long as I could remember, I'd always been focused on some goal or other. Now I had nothing—no book to write, no career to pursue, no plan for the future.

On December 1, 1995, just days after sending off my manuscript, I read that a city council member had resigned and that an aide to Congresswoman Carolyn Maloney named Gifford Miller was running to fill the vacancy. I decided I'd finally take Eric up on his suggestion that I give politics a try. I called up Miller's campaign to volunteer my services and, to my great surprise, Miller himself answered. When could I start, he asked. Right away, I said. Fifteen minutes later I walked to the address he gave me.

From Hollywood films, I imagined his campaign office would be crammed with desks and posters and filled with a cacophony of ringing telephones and shouting voices. Instead, Miller was running his campaign out of his apartment and he had no staff other than an elderly and frail chain-smoking man who ran the local political club. But Miller, a well-spoken twenty-six-year-old graduate of Princeton University, impressed me.

Two opponents quickly emerged: Judith Marcus, a Republican who'd worked for the retiring council member, and Drew Schiff, a Democrat who was a doctor and the son-in-law of Vice President Al Gore. Since this would be a "special election" in which there would be no primary, Miller feared he and Schiff would split the Democratic vote and Marcus would win. His immediate focus, however, was collecting enough signatures to get on the ballot. This was surprisingly hard. First, there were many technicalities designed to trip up outsiders. Signatures had to be witnessed by either a registered voter in the council district or a notary public. If someone wasn't a registered Democrat or had previously signed another candidate's petition, his signature didn't count. Most important, the person had to be a registered voter in the council district, which was challenging since somebody walking down the street on the Upper East Side is as likely to be from Queens or Sweden as from around the corner. Then, right in the middle of the three weeks we had to collect signatures, a massive snowstorm hit and people who are trudging through slush aren't eager to stop and sign petitions.

When I'd volunteered, I figured I'd just be answering phones. Instead, Miller put me in charge of getting him on the ballot. The campaign swallowed me whole. I worked from morning till night seven days a week. Three weeks later, we had more than enough signatures. Schiff, however, did not, so the election was now a straight-up contest between a Democrat and a Republican in a district that leaned Democratic. Miller won with 57 percent of the vote. This was an important lesson in the reality of local elections: they usually turned not on soaring oratory or debates or carefully crafted position papers but on hard work and good organizational skills. I felt good that I'd been able to make a difference.

Miller asked me to be his chief of staff, and while I was considering that offer, I got three others: statewide director of volunteer recruitment for the Clinton-Gore campaign, head of fund-raising for the state Democratic Party, and campaign director for Congresswoman Carolyn Maloney. I was astonished. In academia, you could spend six years slaving away on a doctorate and fail to get a single job offer; after six weeks in politics, I had four.

Eric, I realized, had been right. Politics came naturally to me. Compared to academia, it was like falling off a log. I was good at the practical organization work involved in politics and at getting people to work hard. Moreover, I was excited by the idea that I could use my skills to help people. It reminded me of the satisfaction I'd gotten from helping Cambodian refugees. Given that education was my field, I was particularly interested in improving the public school system. One should either write something worth reading or do something worth writing, Benjamin Franklin once said. Perhaps, I realized, I was better suited to the latter than the former.

Moreover, I'd come up with an idea for running for office myself: challenging my representative on the city council, a Republican named Andrew Eristoff. When I'd floated this idea to Eric, he'd observed that while he was all for my getting into politics, Eristoff was an incumbent from a well-established Upper East Side family that was worth about a half billion dollars. But for some reason this didn't faze me, and I had a plan for winning that I thought just might work. I saw that politics at the local level was a lot about convincing people to vote for you by having direct personal contact with them so I felt I could win if I campaigned really hard for a really long time. Moreover, I'd figured out how to give myself a head start. Congresswoman Maloney was up for reelection a year before Eristoff was. If I accepted her offer to run her campaign, I could meet local donors, volunteers, and activists, while simultaneously helping a well-respected congresswoman get reelected.

I was excited about my plan and new career. A few weeks in politics had given me the happiness and self-confidence that for years had eluded me. However, I was troubled by a personal issue. Eric and I wanted to start a family but I'd been unable to get pregnant. We went to a fertility doctor. He found nothing wrong with us but observed that women had more trouble conceiving as they got older. I didn't realize that was true even for a woman of my age, thirty-two. I was upset that I'd perhaps inadvertently squandered my chance to have children by waiting too long. To improve my odds, I began fertility treatments.

On August 26, 1996, I was having a cup of coffee with someone when Eric suddenly showed up and asked me to come with him right away. We jumped in a cab heading toward his mother's apartment, where Eric's younger brother, Alexi, a film student, still lived. Alexi,

Eric said, had overdosed on heroin. Eric knew that his brother had used heroin but hadn't realized how serious his problem was and hadn't told me about it because his mother had wanted to protect Alexi's privacy. When we arrived at the home of Eric's mother, she confirmed what Eric had feared: Alexi was dead.

I can't even comprehend the pain Eric's parents felt. As for Eric, he soon fell into a deep depression. He suffered from extreme anxiety, couldn't sleep, and took little pleasure in anything. Ordinarily, Eric had a happy-go-lucky adventurous quality so it felt like a different person was inhabiting his body. It was terribly painful for me and I tried my best to help Eric out of his depression, but I also tried to stay focused on my goals. Perhaps that sounds selfish, but I didn't see how my being less happy would make Eric more so.

That fall, Congresswoman Maloney won handily and I convinced the staff I'd recruited for her campaign to work on mine. Among them was Ilana Goldman, an unusually mature and hardworking young woman who served as my campaign manager. In addition, my college roommate, Kathryn Gregorio, agreed to be my campaign treasurer.

When I told my parents about my plans, they looked as if I'd said I was going to join the circus. They didn't know anyone in politics and it struck them as a shady and risky profession. They said, however, that if that's what I wanted to do, they'd support me, and they became quite involved in my campaign, as did Eric's mother.

My obvious disadvantage was in fund-raising. Eristoff was descended on his father's side from a Russian prince and on his mother's side from Henry Phipps, a tycoon who was partners with Andrew Carnegie. Eristoff had spent $300,000 on his last campaign although it was just a six-week special election. While I couldn't hope to match his spending, I wanted to raise enough so I'd at least have a fighting chance. I began calling people off of lists of other candidates' contributors. Nearly everybody turned me down and many of them were nasty about it, but I just forced myself to keep on calling, and every now and then, someone would agree to contribute. One guy whom I asked to give me $1,000 responded, "If you knew who I was, you'd ask for more," so I did and he gave me $3,000.

Another person I called was Michael Bloomberg, who hadn't yet entered politics. He asked to meet with me and asked me many thoughtful questions about education policy, perhaps because he was already thinking of running for mayor. He contributed $1,000 to my campaign.

In total, I made around fifteen thousand cold calls. I refused to take contributions from the real estate industry, however, because Eristoff had attacked his last opponent for doing so. One of the many checks I returned came from Donald Trump, who summoned me to Trump Tower to explain myself. It wasn't personal, I said, and explained my reasoning. Trump was amused. I suspect he felt that a five-foot-two former history professor who refused to take contributions from the real estate industry wasn't likely to go far in politics.

I also called up many voters. I'd spend ten or fifteen minutes on the phone with someone just to get one vote. When I began speaking, the person on the other end would sometimes say "I hate these recorded calls," and I'd have to convince them I wasn't one. My parents and Eric also called voters and we held "meet-the-candidate" events to which people would invite their friends and neighbors. Even if only a few people showed up, they would talk to their friends and neighbors. All of this outreach took an incredible amount of work, but I was blessed with a team of idealistic and energetic staff members whose youth was so striking that a local newspaper printed a cover with caricatures of them sucking on pacifiers.

As Election Day approached, I did more street campaigning, particularly focusing on the rush-hour traffic at subway stations. Senator Chuck Schumer told me to go to the same station in the afternoon at which I'd campaigned in the morning because people would think I'd been there all day and was incredibly hardworking.

On October 19, the *Times* wrote an article about the campaign titled "Money's No Object in This Council Race." While technically an objective news article, it was slanted in my favor (how often do you hear me admit that!). It noted that I was a former professor who'd "assembled a list of campaign contributors that includes dozens of professors and teachers and at least one student (who gave $1)" while Eristoff was a former tax lawyer whose list of contributors included "big donations from the Republican County Committee . . . and occasionally reads like the Social Register, with names like Rockefeller, Frick, Loeb and Luce."

Most people assumed the *Times* would endorse Eristoff, but he had an Achilles' heel: his refusal to participate in the city's campaign finance program. Moreover, I put together a book on my policy positions so that the *Times* would know that I wasn't a typical Democratic politician beholden to the party machine, but rather was willing to take unorthodox positions such as ending tenure for public school principals to make them more accountable.

On October 30, the *Times* came out with its city council endorsements. Eristoff, said the *Times,* had been a good council member, but his refusal to participate in the campaign finance program was "an overriding flaw." "Fortunately," wrote the *Times,* "voters have a good alternative in the Democrat Eva Moskowitz, a history professor who has campaigned hard and given serious thought to the issues facing the Council." Realizing I might actually win, Eristoff doubled down, spending more than $800,000 in total, most of it his own money, a record amount for a council race.

But I felt good about my chances. It seemed like our grassroots campaign efforts were working. When I'd campaign on the street, many people would tell me they'd had contact with my campaign: that they'd spoken to me or to my family on the phone or had been invited to a house party or a fund-raiser by a neighbor or a friend. We prepared for Election Day like it was D-day. My father built hundreds of wooden stanchions for my posters and we convinced hundreds of volunteers to campaign for us. On Election Day, I woke up at five, voted first thing, and then started campaigning. People came up to me in droves telling me they'd voted for me. An elected official who campaigned with me said she'd never seen anything like it and that

I was sure to win. Just before the polls closed, I went home to change before attending the election night party at which I would either declare victory or concede defeat. I rested for a few minutes to get my energy back and Ilana soon called to give me the results.

Eristoff had beat me, albeit narrowly. At first I couldn't believe it. Everyone had worked so hard and so many people had told me they'd voted for me. Yet from the beginning, I recalled, many people had warned me that no matter how good a campaign I ran, I'd never beat a wealthy incumbent. Now, I felt naive and foolish for having imagined otherwise. Moreover, I'd convinced my supporters to join me in this delusion. Now they'd find out that I'd sold them a bill of goods, that all of their hard work had been for nothing. I felt both embarrassed and guilty, and I dreaded facing them. But I did, thanking them profusely for their hard work while vainly fighting back tears.

The following morning was brutal. When you lose an election, your loss is incredibly public. Everyone knows. I was embarrassed and wanted to crawl into a hole. I also felt rudderless. When you've been consumed with something, it's devastating when it suddenly disappears. I nonetheless spent the next several weeks calling all of my donors and volunteers. Several of the more politically sophisticated people I called said something that lifted my spirits. Precisely because I'd been considered a long shot, they said, I'd impressed many people by even coming close. As a result, I was now viewed as an up-and-comer.

But what should I do next? I didn't want to work for another politician because most of them made too many compromises for my taste. Besides, I didn't want to become part of the regular political machine. Instead, I took a job at Prep for Prep, a nonprofit program that worked with gifted minority children.

Because of Eric's depression and my focus on my campaign, we'd stopped fertility treatments, so I was quite surprised to discover in January 1998 that I was pregnant. One day several months later, however, I suddenly began bleeding. I jumped into a cab to go to the nearest hospital and told Eric to meet me there. Although it was obvious

I'd miscarried given how much blood I'd lost, a doctor ordered an ultrasound just to make sure. As they spread cold jelly over my swollen belly, I held Eric's hand fiercely, mourning our lost child. It had taken me four years to get pregnant; who knew if I'd even succeed again, much less carry a child to term? The doctor turned the ultrasound screen toward us. "See this?" she asked, pointing to something that was barely visible but seemed to be blinking. "That is your baby's heartbeat." It seemed impossible after all that blood. I cried tears of joy. It was one of the happiest moments of my life.

I later learned I'd had a hematoma, a condition in which the placenta partially detaches from the uterus and blood fills the void. Often it does cause a miscarriage but not always. Several months later, I gave birth to a healthy baby we named Culver after a relative of Eric's.

By now, Eric had recovered from his depression, and one day at work, he read that New York was about to pass a law authorizing the creation of charter schools. Eric had long been interested in school choice and had come to believe even more strongly in it after noticing that all of the African Americans on Congressman Rangel's staff had gone to parochial schools. This reinforced Eric's sense that public schools weren't providing African Americans with the same level of opportunity that they were providing to white people.

Eric put out the word that he'd offer his services pro bono to anyone who wanted to start a charter school and was eventually put in touch with a woman named Kristin Kearns Jordan whom he helped found the Bronx Preparatory Charter School, on whose board Eric served for many years thereafter.

On June 24, 1999, I learned Eristoff had resigned his seat to join the Giuliani administration. I still yearned to become an elected official but I worried that the demands of politics would prevent me from giving my son the attention he deserved. Eric, however, strongly encouraged me to run anyway. My mom had worked hard, he observed, and I'd turned out fine. Besides, he said, I'd be happier if I pursued my ambitions and that would make me a better mom. I found Eric's support reassuring and decided to run again.

Because I'd done well against Eristoff, no serious Democratic candidate challenged me, but a formidable Republican opponent soon emerged: Reba White Williams, a businesswoman who had impressive credentials including a position on the board of directors of Alliance Capital. Moreover, like Eristoff, she vowed to fund her own campaign generously.

I hired a talented young lawyer by the name of Anessa Karney as my campaign manager and she in turn recruited a hardworking team. I was now particularly glad I'd made such efforts to thank my donors the last time around since I would now be calling on them again for their support. Most were happy to provide it. Then one day a woman who happened to be visiting a nonprofit next door saw my campaign office and walked in. She turned out to be Judy Rubin, a philanthropist and the wife of Robert Rubin, the former secretary of the treasury. She generously offered not only to contribute to my campaign but to hold a fund-raiser for me.

The unions, who usually backed Democratic candidates, were another potential source of support but they made it clear I'd have to "evolve" my positions on several issues such as my opposition to tenure for principals. I wouldn't, and most of the unions, including the UFT, backed Ms. Williams.

Ms. Williams was known for her take-no-prisoners approach to politics. After getting into a feud with Giuliani, she'd offered a $10,000 bounty to any journalist who'd dig up dirt on him. At debates, she belittled me. "This is not a job I need," she boasted, "but for Eva this would be the best job she's ever had." She also criticized me for bringing my son to some meetings of the community board, of which we were both members. "I didn't know who she was," said Williams, "I thought she was just the woman with the baby."[19] While these attacks didn't trouble me, I soon found out I'd made a huge mistake that did. Given that I had a young child and community board meetings went on until all hours of the night, I'd sometimes left before they ended. Unfortunately, that's when most of the votes took place, so I'd missed many of them and Williams sent out a piece attacking me for this. I felt terrible. It raised issues I had about the conflict between pursuing my professional ambitions and being a good mother. I was also incredibly frustrated with myself. By making this one stupid mistake, I'd jeopardized not only all the work I'd done but that of my supporters and family members.

As the election got closer, Ms. Williams promised to spend even more than Eristoff had but we soon got good news. I was endorsed by all of the major papers: the *Times,* the *Daily News,* the *Observer,* and even the *Post,* which rarely backed Democrats. One person who stayed neutral in the race was Donald Trump. He commented, "They're both extraordinary. I'm going to vote for both of them."[20]

On Election Day, we again mounted a full-court press and this time around it worked. We'd won. I was overjoyed that I would finally be able to use my talents to help people. I was also happy not to have disappointed everybody who'd supported me and to whom I was incredibly grateful. I had no idea, however, what a strange institution I was about to join.

23

MOM, WHY DON'T YOU OPEN UP MORE SCHOOLS?

2009–2010

Before Success, I'd never managed more than a dozen employees. By the fall of 2009, Success had ten times that number and a budget of over $20 million. It had turned out to be the most important, engrossing, and challenging thing I'd ever done but, like many women, I felt torn about not spending more time with my children. I wondered what was morally right. At work, I was responsible for educating over a thousand mainly disadvantaged children, while at home I had only three children who were hardly disadvantaged, but they were *my* children and I feared they might resent my absence or, even worse, simply come to view me as some distant figure.

If you're hoping that this is the part of the book where I tell you how women can have it all, you're out of luck. I think that it's just an irresolvable dilemma. My children would have benefited from my having had more time to read to them and talk with them. That they didn't suffer more reflects not some brilliant plan of mine but dumb luck. Many children have learning or physical challenges that require a lot of parental involvement but, for some reason, all three of my children are quite sturdy both psychologically and physically. In addition, Eric is a particularly involved father. And our parents have been very generous with their time, particularly Eric's mother. She often played board games with our children, helped them with their homework, and cooked dinner for them. When they got sick, she'd come over to care for them and take them to the doctor. She was also a specialist in learning disabilities and helped Culver with speech and reading issues he was having. And finally, I was very fortunate to find a wonderful young woman by the name of Teresa Witkowski to help me with childcare. I hired her two weeks after Culver was born and she still works for us today, nearly two decades later. She's like a second mom to our kids.

All of this help allowed me to focus virtually every waking hour on Success. I didn't socialize, have hobbies, watch TV, or exercise. Aside from the time I managed to carve out to spend with my family, I worked all day every day. Success required so much of my time in part because the issues multiplied as we grew. New things arose every day. For example, I learned that one of our deans was dating one of our teachers, which was a problem since deans evaluate teachers. I asked Eric to draft what is known as a "cupid policy." It required employees to disclose certain romances (such as the dean-teacher romance) so we could address any issues it created. It also prohibited romances between an employee and someone in their direct chain of command, such as a principal and a teacher in the same school. There was only one person, Eric explained, who couldn't become romantically involved with any other Success employee: me, since everyone was in my chain of command. "Funny how that worked out," he added.

Another issue arose when a parent told us that her child had been slapped on the head by one of our teachers and then the parent of another child made a similar complaint about this teacher. I was conflicted about what to do. This teacher had been with us since our first year, was totally committed to teaching, and really seemed to reach her students, whom she'd motivate by doing things like playing the *Rocky* theme song. She denied hitting the kids, but she must have had a rough style given that there had been two complaints, and since she wasn't acknowledging she was doing anything wrong, I didn't see how the situation could improve. I decided I had to let her go. It was a painful decision, and I'm not sure I got it right. Sometimes you just can't get to the bottom of things so you're left with uncertainty and, knowing full well you may be wrong, you just have to use your best judgment.

This incident made us realize, however, that we needed to give our teachers clearer guidelines about physical interactions with students. Sure, everybody knows you can't hit students, but there are more gray areas than you may think. Suppose, for example, that a boy in kindergarten is taking all of the books off the shelves and throwing them on the floor. It's natural to want to grab his arm to stop him but if the child resists, he may get hurt. If that happens, the parents will probably blame you. Here's an email I got from one of my principals about one incident:

> [A student] bit me in the hand hard enough to break the skin and draw blood. Unfortunately, my reflex was to slap the young man on the forehead to get him off my hand. I informed the parent and apologized but said there wasn't much I could do as he was biting my hand at the time. The mother contacted the police and they sent a team of officers over to question me. When I showed them the bite mark on my hand, they informed me that they were throwing out the complaint.

Given the dangers in this area, we developed strict policies. Our teachers are never allowed to use force with a child—even grabbing a child's arm—unless he's acting dangerously.

These types of issues multiplied, but one advantage of having a larger organization was that we could now afford to hire more people to help me manage our schools. We had created a separate not-for-profit organization called the Network that was responsible for hiring principals, recruiting teachers, renovating facilities, fund-raising, financial management, and professional development. These were all things that could be done better and more efficiently through centralization. After a lengthy search, I hired Keri Hoyt, who'd been with *The Princeton Review* for seventeen years, to help me run the Network. She turned out to be a godsend. As she was both decisive and hardworking, she could plow through enormous amounts of work at an astonishing clip. She also helped us move to a new space. We'd been managing all of our schools out of a couple of rooms in PS 149, where I didn't have an office or even a desk, just a stool I'd sit on while I wrote emails and made phone calls. Eventually even my stool disappeared and I became nomadic sitting on the main office's couch or the desk of someone who happened to be out. We managed to find offices at 125th Street and Lenox Avenue, Harlem's epicenter, and I actually had my own office, which seemed like an incredible luxury.

Another advantage of running several schools is that my principals could learn from one another. For example, one of them, Jim Manly, found that his kindergarten teachers were getting frustrated because their classrooms were chaotic. The problem, he realized, was that the teachers hadn't invested enough time in helping their students learn the routines that make a classroom work. The following year, he had his teachers spend much of the first week practicing routines. After recess, for example, students practiced going into the classroom, putting away their coats, and sitting at their desks. If they didn't do it right, they'd put their coats back on, line up, and do it over again. While the teachers preferred to get to the substantive instruction as soon as possible, Jim found investing time up front in learning routines led to more progress and less frustration in the long run, and his approach soon spread to our other schools.

Some people think we are too obsessed with order, but it doesn't stem from some preconceived ideology. Rather, it is based on experience. We've seen firsthand the difference between a school whose teachers make this investment in teaching management and routines and a school that doesn't.

While we were obviously serious about academics, we also believed school should be joyous and engaging. Every year, I took my own children to the Big Apple Circus, an old-fashioned one-ring circus, so I decided to take all of our students in kindergarten through second grade and hundreds of their parents to this circus. The children were reduced to stitches by the great clowns Bello and Grandma and were amazed by the acrobats from Italy, contortionists from China, equestrians from Kazakhstan, and a man from Spain who "juggled" five Ping-Pong balls with his mouth. We also took our kids to zoos, museums, farms, and the Paper Bag Players, a group of brilliant performers I'd enjoyed seeing as a child. Eventually, we realized it would also be good to bring performers into the school, so we brought in a juggler, a magician, and a remarkable marionettist I'd seen one day when I was walking in Central Park with my kids.

Being co-located often presented us with challenges. One day, one of our students found a gun in Harlem 3's building. We tried to find the principal of the co-located school, but, as usual, he was missing in action. DOE investigated and discovered he was regularly visiting his mistress during school hours. We also had trouble getting space to store books and supplies. We were always told the closets were full, but when I'd insist they be opened, I'd find dusty textbooks that hadn't been used in decades. At one of our schools, we found a storage room in the basement that was empty except for a couch, a lamp, and a few personal items such as clothes, which the custodian removed at our request. A few days later, an elderly disheveled woman accosted me, saying she was a former teacher, that the co-located school's former principal had agreed to let her live in the basement, and that I'd thrown out her belongings. I'd actually gotten occasional reports about an elderly woman who wandered around the schools but I'd had no idea she lived there.

As we approached the end of our third school year, we had to figure out which teachers to keep. Most principals find it hard to let teachers go. They're inclined to decide that a mediocre teacher is good enough or give her one more chance. Of course, it's natural to feel compassion for a weak teacher, particularly one who's really trying but just isn't suited to the profession, but giving a teacher "one more chance" means thirty children will likely get an inadequate education for a year at a critical stage of their development. Principals sometimes unintentionally put the needs of teachers first because they are closer emotionally to the teachers than to the students. Since my own children attended Success, I was particularly aware of the impact that keeping a mediocre teacher had on kids. I wrote to my principals:

> We are not an employment agency. [You] cannot let your attachment to individuals override what is good for children. I think about it this way. If I would prefer for Hannah and Dillon NOT to have a certain teacher, I cannot in good conscience bring them back and have other people's children suffer.

That summer, a documentary about Success called *The Lottery* came out. It brilliantly weaved together our fight to get space for new schools with the stories of four families whose children were in our lottery. We arranged to screen this film for our families at the famous Apollo Theater. I took Culver, and when he learned at the end that some of the children in the film hadn't won the lottery, he cried. "Mom," he asked, "why don't you open up more schools?" This comment was a deep relief for me since it meant that Culver understood the importance of my work and didn't resent me for spending so much time on it.

That summer, we also got our test scores. I was concerned about how well our students would do because SED had announced it was making the tests harder. Sure enough, passage rates plummeted citywide to 54 percent in math and 42 percent in English. Our students, however, continued to perform strongly: 95 percent passed in math and 88 percent in English.

These strong test scores left me optimistic for the future. What I did not know, however, was that the 2010–2011 school year would see not only an existential threat to Success from our traditional enemies but the addition of a new one, a storied civil rights organization that would accuse us of being "slave masters."

24

THURGOOD MARSHALL MUST BE SHAKING HIS HEAD

2010–2011

In 2010, 59 percent of Manhattan residents were white, but more than two-thirds of its district schools had fewer than 10 percent white students. The district school system's design encourages segregation. First, elementary schools have small zones, so that even where housing projects are near expensive condominiums, the children living in them rarely attend the same schools. In some cases, this is intentional. For example, residents of Lincoln Towers, a private development, got the city to create a new school for them that was zoned to exclude children in a public housing project just two blocks away despite a lawsuit that correctly observed that this plan would cause segregation. Where affluent residents are zoned for schools with poorer students, the school system offers them escape hatches such as gifted and talented or dual language programs. Moreover, while it should be easier to integrate middle and high schools because older students can travel farther, the most desirable schools at this level have admissions criteria that children who've attended a failing elementary school rarely meet.

There's no malevolence behind this system. Rather, it's the unintended result of affluent families' efforts to ensure their own children are well educated. They rent pricey apartments or purchase homes zoned for the best schools, hire tutors to help their children get into gifted and talented programs, and find out about the best new magnet schools by the application deadline. Through these efforts, they snap up the spots at the best schools and disadvantaged families are left with the worst ones.

In the fall of 2010, virtually all of Success's 2,400 students were children of color. While we were giving these students an excellent education, I nonetheless felt we should do our part to diminish segregation by making a deliberate effort to attract a racially and socioeconomically diverse student body at our next school. We found the perfect site for such a school: a building on Manhattan's Upper West Side that was just blocks away from both low-income housing and multimillion-dollar apartments. Unfortunately, the school this building contained, PS 145, was too academically weak to attract affluent families.

Bill Perkins should have been delighted we were opening a school on the Upper West Side because he'd complained that "charter schools . . . are only being put in communities of color." On October 19, however, he helped organize a protest at which he warned Upper West Siders that sharing space with a charter school "niggerizes you, it makes you second class."[21] Council Member Gale Brewer vowed to "strangle any parent I find who moved [their child] into a charter school." (A couple of years later, Brewer would comment, "I don't like Eva Moskowitz, I'll be honest . . . I'll tell her that, right to her face."[22] How does one even respond to that?) Rounding out the field was Noah Gotbaum, who was head of the Community Education Council and an aspiring politician who would soon bear the unique distinction of having his own stepmother endorse his opponent. With such an august assemblage leading the charge, Juan González couldn't be far behind. He argued that Success shouldn't be allowed to co-locate with PS 145 because, while underenrolled, PS 145 had "blossomed in recent years" and was sure to "grow under [a] new federal grant."

The following day, the Community Education Council held a hearing on our co-location that the *Daily News* accurately described as "virtual mob rule." I had to attend our annual fund-raiser that night, but I sent one of our staff members, Larisa Beachy, to videotape the meeting so we'd have a record. Gotbaum didn't like this so he had Larisa arrested and hauled off in handcuffs to the cheers of the crowd. I was livid. Larisa had a right under the New York Open Meetings Law to videotape this public meeting. Moreover, I felt terrible that Larisa, a young woman who'd taken this job to help disadvantaged children, had been subjected to this humiliation. However, I was proud of her for standing her ground. As Eleanor Roosevelt once said, "A woman is like a tea bag; you never know how strong it is until it's in hot water." When Larisa emailed her father, a former mayor of New London, Connecticut, about her arrest, he responded proudly, "You GO!!! girl." However, the principal of the school where Larisa worked emailed me that "I want to make sure that the risks we ask people to take are reasonable and worthwhile. . . . [M]y father hat came on last night thinking about poor Larisa in handcuffs." I replied: "Look at all the risks people have taken for justice, human rights, First Amendment rights. . . . The parent in us needs to be proud [because] there are principles worth standing up for."

After this incident, DOE added injury to insult by withdrawing our co-location. González had predicted that if only we were prevented from co-locating with PS 145, it would thrive and grow. In fact, five years later, its enrollment had shrunk by 37 percent, and its English passage rate had fallen to 15 percent. To this day, parents in this neighborhood still don't have a good option for their kids since we weren't allowed to open up there. As for the federal grant González mentioned, perhaps you'd like to know how your 11.5 million taxpayer dollars were spent. The grant was supposed to diminish segregation by increasing white enrollment at primarily minority schools like PS 145 and increasing minority enrollment at primarily white schools like PS 87. In the end, both PS 87 and PS 145 ended up more segregated than they started—and no, you're not getting your $11.5 million back.

In the middle of all this, Chancellor Joel Klein called to tell me he was resigning and asked me if I wanted to succeed him (although he didn't say whether Bloomberg was on board with this). I decided, however, that I couldn't abandon Success. I was sad to see Klein go. He found it deeply painful that other children weren't getting the same quality public school education he'd had, and his commitment to rectifying that had led him to make enormous financial sacrifice and endure vitriolic opposition.

DOE proposed another co-location site, PS 165, but again protests arose. As the *Times* noted: "Members of the teachers' union and New York Communities for Change, which replaced . . . ACORN, are often present at rallies and copied on e-mails debating the next steps in the battle." DOE withdrew its proposal and I called Deputy Chancellor Marc Sternberg in frustration to tell him he needed to identify a location and stand his ground. I'd take anything, I said. "Fine," he replied sarcastically, "then you can have Brandeis," which Sternberg probably figured was a nonstarter since it was a high school building with some tough kids. I figured, however, that even the toughest high school kids weren't going to pick fights with kindergartners so I said, "Great!"

Again, there was opposition. "We don't need more options here," said Gotbaum. "We have options. We have great schools." Perhaps that was true for him, since he sent his children to PS 87, which raised tons of money through private donations (as well as $1.3 million in federal funds for enrolling more kids of color but not actually doing so). The options weren't so great, however, for families who didn't have a couple of million dollars to plop down on an apartment zoned for PS 87.

Despite this opposition, however, DOE stuck to its guns and, on February 1, formally approved our location. Two weeks later, however, Jenny Sedlis, our head of external affairs, got an email out of the blue from an attorney who'd heard a lawsuit was going to be brought to challenge our co-location. This attorney, Emily Kim, had put her son in our lottery, and she offered to represent us for free if a lawsuit was brought. Jenny called Eric to ask if he'd heard of this woman's firm, Arnold & Porter. Eric laughed; it was one of the best in the country. It was like Michael Jordan strolling by and asking if he could get into a pickup game. Eric went to meet with Kim. Ostensibly, it was about this case, but we had another agenda as well: finding a general counsel for Success. Eric was fulfilling this function as a part-time pro bono lawyer but Success was getting too big for that. It turned out that, in addition to being an excellent lawyer, Kim had previously taught in Africa and helped kids with special needs. Eric felt she might be the one. But first we figured we'd see how she did representing us pro bono.

On April 8, the lawsuit Emily had warned us about was filed. The plaintiffs claimed that our elementary school students might sneak weapons into the building and also that, since our school would have "geographic diversity," e.g., kids from Harlem, transporting them to school would contribute to pollution. Opposing integration on environmental grounds is, I guess, the politically correct form of bigotry.

The lawsuit also advanced many hyper-technical objections to the approval process: that a hearing notice distributed forty-five days in advance had omitted the hearing date, an omission corrected two days later; and that it had taken five days to translate the notice into Spanish, although that was still forty days before the hearing. Not one single person claimed, however, that these mistakes had actually resulted in their missing the hearing on our co-location.

Then, on May 18, a far bigger bombshell dropped. The UFT and NAACP brought a suit to invalidate nineteen charter school co-locations, including seven of ours, because the city had allegedly made mistakes in the "Educational Impact Statements" and "Building Utilization Plans" it had issued. These documents required that the city specify precisely how many rooms each school would get and set a schedule for sharing certain facilities such as the gym and cafeteria. The UFT and NAACP claimed we'd been given a disproportionate amount of space, citing examples of what seemed on the surface like unjust disparities: we got exclusive use of the cafeteria for breakfast at one school, while the three district schools shared it for theirs; we got the "big gym" at one school, while the district school got the "small gym"; we were allocated gym time at another school while a co-located school for special needs kids wasn't. But in every case, there were simple explanations: our students ate breakfast alone because our school started earlier; the "big gym" held 369 students, just 9 more than the small gym, which held 360; the special needs school hadn't been assigned gym time because that school's principal said he didn't want his kids using the gym which, whether or not that was a good idea, certainly wasn't our fault.

Moreover, the UFT and the NAACP weren't asking that these inequities be fixed: that we use the small gym rather than the big gym, or that gym time be allocated to the school with special needs kids. Rather, they wanted the courts to prohibit us from using these buildings *at all*. Why? Because their goal wasn't to fix inequities but to use the appearance of inequities to stop the spread of charters.

But the fact that these claims were ridiculous by no means meant we'd prevail. In one case, Girls Prep, a charter school that Eric had founded, had lost its co-location because DOE's impact statement had failed to mention that a small school in a building with several schools was going to lose a couple of rooms it didn't need. The co-location was annulled and Girls Prep had to rent private space for $1 million, but the rooms Girls Prep was going to use just sat empty. Thus, a school that served poor girls had been forced to stuff a million dollars down a rat hole because DOE had made a technical error.

In another case, the UFT had successfully sued to prevent DOE from closing seventeen schools. Even if these schools deserved to be closed, the court had ruled, DOE couldn't do so because its educational impact statements hadn't listed every single extracurricular program at these schools and similar programs at other schools. Moreover, DOE couldn't correct these errors because school closures had to be approved six months in advance. Thus, when the city made a mistake, it was game over for the coming year. None of this was accidental; the law was intended to slow down school closures and charter growth.

This decision illustrates why the government has such trouble running good schools. Bloomberg thought the best solution to failing schools was to close them while the UFT and its allies thought the best solution was to fix them. The lawsuit, however, didn't resolve that conflict since it didn't permanently prohibit Bloomberg from closing these schools. Rather, it merely required him to keep them on death row for another year, a bad outcome no matter what view you took on school closings. This type of tug-of-war is inherent in the political process and just isn't a good way to run schools.

Thus, there were some bad precedents for us and the NAACP's participation in this lawsuit lent it moral credibility. Why, you ask, was the NAACP involved given that we served primarily African American kids whose families desperately wanted them to attend Success? First, the UFT gave money to the NAACP. Second, the NAACP's membership included many unionized teachers who wanted to protect the district schools, and comparatively few younger members with school-age children concerned about improving the quality of the public schools.

This lawsuit illustrated the UFT's tactical brilliance and long-term thinking. The UFT had managed to create this fiction that sharing space with a charter school was harming district schools with rent-a-mob protests and the help of ideological accomplices like González and politicians the UFT had supported. Then the UFT had used the fiction it had invented to support legislation it had drafted that was brilliantly disguised as good-government transparency but in fact could be deployed to tie up charter schools in red tape, and now it was springing this trap that it had been been carefully constructing for years and was doing so with moral credibility it was renting from the NAACP.

I'm used to high-stakes poker and don't scare easily, but this lawsuit frightened me. If we lost, we'd have to shut down most of our schools, abandon hundreds of students who would end up in dysfunctional district schools, and lay off half of our teachers. I wracked my brain to think of a contingency plan. I considered finding alternative facilities, but existing school buildings weren't just lying around waiting to be used. Moreover, we had neither the time nor the money to turn a non-school building into a school, which is quite an undertaking given the stringent safety and architectural requirements for schools. Eventually, I just gave up on figuring out a contingency plan. There wasn't one. Losing meant disaster.

I set about figuring out how to win the lawsuit. Judges claim they base their decisions only on the law, but I believe public opinion matters in a case like this. For that reason, I felt I couldn't ignore the NAACP's involvement, that we had to protest the NAACP. My staff thought I was nuts. Who protests the NAACP? I felt, however, we had to make clear that the NAACP's views were at odds with those of the families in Harlem the NAACP purported to represent. I wanted to make it clear, however, that we weren't against the NAACP, just their position on this one issue, so we decided to have our students wear T-shirts that said Future NAACP Member. This reflected our belief that the NAACP's opposition to charter schools was backward-looking and that the NAACP would in time come to support charter schools.

On May 26, thousands of children wearing "Future NAACP Member" T-shirts and parents carrying signs that read "NAACP: Don't Divide Us, Unite Us" marched to the plaza at 125th Street and Adam Clayton Powell Jr. Boulevard.

The speakers communicated their outrage and despair at the NAACP's actions. "I could barely believe my ears," said one of our teachers, "when I found out that the NAACP was trying to shut down some of the most successful schools serving black and brown children in our city." A parent said:

> My child cannot be told that she's not going to get to go to her school in September. I cannot look her in the eye, as a parent, and tell her, "Well, the problem is that this group of people that Mommy told you about during Black History Month, that did all those great things a long time ago—they want to stop you from doing great things . . ."
>
> NAACP, please, don't turn your back on my little girl. Turn your back on this lawsuit instead.

Thousands of African American parents protesting the NAACP was a man-bites-dog story that drew considerable attention. Many prominent African American leaders bemoaned the NAACP's actions. Kevin Chavous observed that "in the heavens above, legendary figures such as Thurgood Marshall, Walter White and Roy Wilkins must be shaking their heads" to see the NAACP become "the target of a protest by the people it was created to serve." The NAACP "seems to have switched sides," wrote United Negro College Fund president Michael L. Lomax. "It's fighting not for the right of kids of color to get a good education, but to keep failing public schools open and to limit kids' ability to go to public schools that are working." The NAACP's "unholy alliance" with the UFT, wrote *Daily News* columnist Stanley Crouch, was "proof of how low a great civil rights organization has fallen since its days of advocating for racial equality."

In response to the rally, the NAACP claimed we'd "unfairly singled" them out. Really? They'd sued to evict our schools and *we* were singling *them* out? Hazel Dukes, the NAACP's New York leader, commented that parents "can march and have rallies all day long" but "[w]e will not respond." It was startling to hear the NAACP proudly declare its indifference to heartfelt protest by the very people whose interests it purported to champion.

The NAACP and UFT responded with a counterprotest in front of our offices. It was small and consisted entirely of politicians, union officials, and paid staff. Dukes spoke first. "I live here in Harlem," she said repeatedly. This was the thinly veiled "you don't belong here" charge to which I'd been subjected by City Council Member Arroyo. She then made some vacuous remarks about being for "all" kids but failed to explain how this could be squared with the fact that her actions would hurt our kids. Perkins claimed that "Co-location is a form of segregation [because] you put a wall between one kid and another kid." It seemed to me, however, that putting us in a whole other building rather than co-locating us would make us more separate, not less. Dukes soon called an African American founder of another charter school a "dumbass," telling him, "You went to Harvard on my back, not because you're smart, not because you worked hard."

One of our parents emailed Dukes:

> Due to this wonderful school, my daughter can read. . . . If you . . . continue on this horrible lawsuit against my daughter's school and the fellow eighteen charter schools, it will not be the best legacy to leave behind . . .

Dukes responded: "You are not a member of the NAACP and don't understand that you are doing the business of slave masters."

Seeing that things were getting out of hand, Ben Jealous, head of the NAACP, went into damage control mode. They'd "just had no idea," he claimed, "how much tension [the lawsuit] would create." Of course, who could possibly predict anyone would get in a tizzy over a little thing like a lawsuit to evict nineteen schools? Jealous claimed the NAACP was "open to all options to settle this suit." That was news to our ears because we would have been happy to exchange the infinitesimally larger "big gym" for the "small gym" and assign gym time to the special needs students, but we'd have to see whether Jealous would really stand up to the UFT to advocate for such a compromise.

The NAACP and the UFT did let Geoffrey Canada's schools out of the lawsuit, recognizing the stupidity of taking on the most prominent African American educator in the country. While I was happy for Canada's schools, I was distressed that hundreds of students for whom I was responsible might suffer just because I didn't qualify for this favorable treatment.

Fortunately, the great law firm Paul, Weiss agreed to represent us pro bono. There were two legal issues. The first was whether space in the co-located schools had been allocated fairly. On that point, we felt good because, well, it had been. But there was another issue: whether the Building Utilization Plans (BUPs) had adequately explained why the allocation of space was fair. For example, we were scheduled to use the cafeteria alone for breakfast because our school day began earlier. But, said the UFT and the NAACP, it wasn't enough to explain that *now,* you had to have explained that *in the BUP itself.*

So why not just revise the BUPs? That's exactly what the city wanted to do, but we figured the UFT and the NAACP would argue that the BUPs couldn't be revised after the co-location was approved. The city went ahead anyway, allocating gym time to the special education school and explaining why our students ate breakfast alone and how the big gym was virtually the same size as the small gym.

Just as we predicted, the UFT and the NAACP claimed it was too late to revise the BUPs to address the inequities identified in the lawsuit, that the court's only option was to annul our co-locations. The NAACP's duplicity in taking this position was truly astounding given Jealous's claim that he was "eager to discuss all options for redressing the inequities . . . and ensuring all kids can attend their school of choice this fall."

With the issues fully briefed, our fate was now in the hands of the court. On July 21, 2011, the court ruled. We'd won! The BUP modifications were legal, the court ruled, and had addressed the issues the NAACP and the UFT had identified. Then on August 13, the court issued another decision dismissing the lawsuit against Upper West as well.

I was incredibly relieved. These lawsuits had weighed upon me more than anything else in my professional career. Losing them would have been far worse than the two elections I'd lost because it would have affected thousands of students and their families.

Two weeks later, Upper West opened. Just as we'd hoped, it was racially and economically integrated: a third of the students were white, a third were Latino, a quarter were African American, and the remainder were Asian and multiracial. Among them was Emily Kim's son. Moreover, since she'd become one of the city's leading authorities on charter school law and had been a fearsome and tireless advocate for us in the litigations, we offered her the job of general counsel, which she accepted.

When families in our Harlem schools learned we'd opened up

a school on the Upper West Side, many immediately suspected it must be better since their experience in the district schools was that people of color always got the short end of the stick. To address their concerns, I had a delegation of parents from our Harlem schools visit our Upper West Side school. What they found surprised them. Everything—the uniforms, the curriculum, the teaching methods, the disciplinary policies—was exactly the same. That was because, from the very beginning, we'd designed Success not for a particular race or class or ethnic group but for all students.

25

PRAYING WHILE RUNNING

2000–2001

Henry Stern, the city's brilliant and quirky parks commissioner who'd graduated from the Bronx High School of Science at fifteen and from Harvard Law School at twenty-two, gave everyone with whom he dealt a parks department badge emblazoned with a nickname he'd selected for you. Mine, I was horrified to learn, would be "Buffy," which sounded to me like a name for an airheaded bimbo. Stern told me, however, that it was actually a compliment: he disliked Reba White Williams, so for defeating her, he'd dubbed me Buffy the Vampire Slayer. However, as I got to know more about the council, I wondered at times whether the vampire hadn't gotten the better of the bargain.

I joined the council at a time of transition. The city's charter had recently been rewritten to give the council more power, but many of my colleagues had been elected before this happened, when the council tended to attract weak candidates. Fortunately, however, the council had a strong leader: Speaker Peter Vallone, a devout Catholic from Queens, a man of considerable integrity and common sense who was underestimated by the media because he had an old-school working-class air about him.

Backbenchers like me had little power on the council. The grown-ups—Vallone, the mayor, and other politically influential players such as the heads of the Democratic county committees—would hash out legislative deals and then order us to approve them. At no time was our lowly status more apparent than during budget season. The city's budget, at $39 billion, was the fourth largest in the nation after those of the federal government, California, and New York state, but most council members had little influence over it. Rather, much as one jingles a key chain before a crying baby, Vallone would give us a few hundred thousand dollars to allocate to nonprofits in our district to distract us from our lack of input into how the remaining $38.99 billion dollars would be spent. My colleagues would then waste their time trading these sums like baseball cards. A council member would offer to match my $10,000 contribution to my favorite nonprofit if I'd match his $10,000 contribution to his, although we'd reach precisely the same result if we each gave our own favorite nonprofits $20,000 each.

As one of the council's newest members, I had even less influence than most, so I had to find creative ways to get things done. Sometimes I did this by bridging the ideological gap between my colleagues, who leaned left, and the Giuliani administration, which leaned right. For example, Giuliani had set aside $20 million for tuition vouchers but the council wouldn't allow this. I figured maybe we could compromise by spending the money on charter schools. My proposed compromise was accepted and I found it both surprising and exhilarating that charter schools were going to get $20 million just because I'd had an idea and made a few phone calls.

While I supported charter schools, I still saw them largely as a sideshow and my principal focus was on fixing the district schools. I discovered, for example, that the escalators at the School of Art and Design, the premier graphic arts school in the city, had been broken for three years, so I got them fixed after a year of relentless effort. At PS 6, my alma mater, I helped get a library built in record time by busting through many bureaucratic hurdles.

Since my powers as a junior member of the council were modest, I focused most of my energy on solving constituent complaints. I noticed, however, that my staff's efforts were often unavailing. We'd write a letter and follow up with multiple phone calls and nothing would happen. Henceforth, I told my staff, we'd write three follow-up emails, so we'd have a paper trail, and then go to the press if we didn't have a satisfactory response. Over time, the departments figured out it was easier to solve our constituent problems than deal with the press.

I liked working on constituent complaints both because it was concrete—you knew you'd helped someone—and also because it gave me a window into problems with the city's bureaucracy that were affecting people throughout the city. I learned, for example, that students at a new school in my district, the Life Sciences Secondary School, hadn't gotten their textbooks a month into the school year. When I asked DOE about it, they told me not to worry my pretty little head because the books had been ordered and would arrive "soon." The students, however, told me the same thing had happened the prior year and I wanted to get to the bottom of the problem so it didn't recur. I asked to see the purchase orders to find out if the books had been ordered late. When the DOE refused to give them to me, I issued a press release charging the department with a "culture of secrecy." This was true, but I also knew it was the type of incendiary language that could help draw press attention. Both the *Times* and the *Daily News* wrote stories and DOE promised to "respond to Ms. Moskowitz's letter and her allegations promptly" and also ordered all of the city's "principals . . . to go class to class and find out what textbooks were missing and why."

The following spring, parents at this same school told me that they feared their children would fail the statewide chemistry exam because they'd had a string of five incompetent chemistry teachers. I again went to the newspapers and this time was able to get DOE to pay for tutoring, but while this enabled some students to pass, for many, it was too little too late.

Another common constituent complaint was the absence of an academically rigorous public high school on the Upper East Side. DOE claimed there wasn't a site available, so I began looking and found a building that the auction house Sotheby's had just vacated. DOE agreed to open the Eleanor Roosevelt High School there. (Curiously, a real estate agent later brought a lawsuit claiming I'd unfairly deprived her of a commission by functioning as an unpaid broker for the site.)

While solving these complaints proved satisfying, I was convinced these problems were systemic, that there were other schools at which textbooks were arriving late and where teachers were incompetent. In theory, the council's Education Committee, on which I sat, was supposed to look into these issues but it wasn't very effective. The committee's chair was often late, so I'd find myself making frantic phone calls to track her down while witnesses we'd invited to testify cooled their heels. When the hearings did start, council members would focus on problems at particular schools in their district rather than exploring the systemic issues that were causing these problems.

While the council's dysfunction was quite dispiriting, I was hoping things would improve in 2002, when term limits would force thirty-six of the council's fifty-one members out of office. *Post* columnist Jack Newfield expressed the sentiments of many New Yorkers regarding this development when he wrote, "I am against capital punishment—except in the case of Adolf Eichmann. And I am against term limits—except in the case of the City Council." On February 6, 2001, the council members who were going to be forced out introduced a bill to overturn term limits. Vallone opposed the bill and my colleagues usually did what he said, but this concerned something on which they had unusually strong convictions: their own political careers. The bill was referred to the Government Operations Committee, of which I was a member, and the vote was set for March 15. Of the committee's nine members, five, including me, voted to keep term limits.

Since Vallone was term limited, this meant we'd have a new speaker the following year. Gifford Miller asked me to support him for this position. Backroom politics weren't my strength so I had little ability to handicap his chances, but I agreed to support him since I felt he'd be a good speaker. Gifford also managed to recruit another colleague of ours, Christine Quinn, and he explained his plan to us. Rather than wait for the new council members to get elected to ask for their support, Gifford wanted to ask candidates for their support now and help them get elected if they agreed to provide it.

In the months that followed, the three of us spent countless hours together and developed a real camaraderie despite our very different backgrounds. Gifford had grown up on Fifth Avenue in the lap of luxury and gone to the very best schools: St. Bernard's, an elite private school; Middlesex, a boarding school; and Princeton. But he came from the kind of wealthy family that had a strong sense of public service. His father had served in the Johnson and Kennedy administrations and his mother had revived a large formal garden in Central Park called the Conservatory Garden.

Chris's maternal grandmother had emigrated from Ireland as a third-class passenger aboard the *Titanic,* whose sinking she'd survived by pushing past the crew who'd tried to keep her in steerage. "When the other girls dropped to their knees to pray," Chris's grandmother told her, "I took a run for it." The moral of this story, Chris once told a priest, was that her grandmother knew there was a time for praying and a time for running. The priest, however, had another interpretation. "Your grandmother," he said, "knew you could pray *while* running." Chris's father was a union shop steward and Chris herself had attended parochial school and then Trinity College, after which she'd become a community organizer and then entered politics as an aide.

As a Jew who'd attended public school, I came from yet another background, but despite these differences, we shared a common vision. A few years earlier, the *Times* had observed that the council was "an unpredictable place of raucous, chaotic and sometimes absurd debates, reflecting a governing body still learning to handle its new role." We wanted to professionalize the council, to make it worthy of exercising its recently acquired powers.

The task before us was to help elect the dozen or so candidates who'd agreed to support Gifford. Most were inexperienced, so our job was to guide them through the process of collecting signatures to get on the ballot, street campaigning, showing up to events on time, handing out literature, and raising money—all of the organizational work you need to succeed in local politics.

While I was excited by this new chapter of my political career, one personal issue troubled me. I wanted another child but had been unable to get pregnant again. Eric and I went back to a fertility specialist and, after a couple of expensive in vitro fertilizations, I became pregnant, but then miscarried. I was devastated. The prospects for my having another child were increasingly small. In the six years that Eric and I had been trying to have children, we'd succeeded only once and, at the age of thirty-five, my odds were only getting slimmer. I decided at least for the moment to refrain from further medical procedures and hope that, despite the long odds, I'd become pregnant naturally.

I turned my attention back to helping Gifford become speaker. Campaigning all over the city led to many strange adventures. When I went to meet a candidate named James Davis at his home, only his mother was there, and while I waited for him to arrive, I smelled something burning. It turned out that his mother had forgotten something on the stove, causing a fire in the kitchen. When Davis finally arrived, I was on the front stoop and the firemen who'd put out the fire were emerging from his home.

Another candidate we supported was a former police officer named Hiram Monserrate. Everything seemed to be going pretty well until Monserrate was arrested on the eve of the election for defacing his opponent's campaign posters. We had to bail him out to ensure that he'd actually be able to campaign on Election Day.

Since I was also on the ballot, I spent Election Day campaigning in my own district. It was a beautiful September day: sunny, cool, and cloudless. I was handing out literature and greeting voters on Sixteenth Street and First Avenue when I noticed an airplane flying unusually low. Suddenly, it slammed into one of the twin towers and thick black smoke began pouring out of the building. People gathered around me, watching in horror. Then there was an explosion from the other tower. Black smoke was now pouring from both towers and we soon got word it had been another plane and that the Pentagon was also being attacked. Then, as we looked on, one of the towers began crumbling, eliciting gasps of surprise and cries of horror since we knew the building contained thousands of our fellow citizens whose lives were being extinguished before our eyes.

I tried reaching Eric, but the cell phone lines were overwhelmed. I soon learned that the election had been canceled and, as the subways weren't working, I began walking home. Thousands of other New Yorkers were also pouring out of buildings and walking home as F-15 fighter planes flew overhead. It was surreal and horrifying: the images of the buildings falling that we couldn't get out of our heads, the billowing clouds of dust, the people walking uptown with worried and bewildered expressions on their faces, the busy signals on our phones. It felt like the city had been transformed into a war zone.

Eric and I decided to get away from the city for the weekend. From a distance, it felt like 9/11 had perhaps just been a terrible nightmare, but when we returned, we saw that the twin towers, which had risen majestically above all of the city's other skyscrapers, were missing from the skyline, which brought back home to us the reality of what had happened.

In the weeks that followed, the city went through a period of mourning. For weeks on end, the *Times* ran a series called "Portraits of Grief" that contained photographs and brief remembrances of most of the more than three thousand New Yorkers who'd lost their lives. Everyone, it seemed, knew someone who'd perished.

In the wake of the attacks, it seemed strange just to go on with the lives we'd led before and yet, of course, we had to go on, which in my case meant returning to campaigning. I did so, and won my primary easily. In the mayoral race, Bloomberg won the Republican primary, and Mark Green, a fairly liberal candidate who'd begun his career working for Ralph Nader, won the Democratic primary. Despite his wealth, Bloomberg was considered a long shot in the general election because Democrats outnumbered Republicans by six to one, but a perfect storm was in the making. Many New Yorkers were worried about the impact 9/11 would have on our city's economy. In the short term, the attack had harmed fourteen thousand downtown businesses. In the longer term, 9/11 threatened to fundamentally undermine the city's financial sector. Financial firms had already been moving their back-office operations out of the city in search of lower taxes and cheaper labor. Now, they worried about concentrating critical operations in a city that was a target for international terrorists. Given these concerns, electing a businessman who'd made his fortune on Wall Street, the very industry most threatened by 9/11, was an increasingly attractive proposition. But even then, Bloomberg still would not have won had Green not made the bizarre decision to endorse a proposal to let Giuliani extend his term by three months so he could deal with the aftermath of 9/11. It was tantamount to an admission by Green that he wasn't up to leading the city in times of crisis. On Election Day, New Yorkers agreed and chose Bloomberg.

Most of the dozen or so council candidates who'd agreed to support Gifford for speaker prevailed, and Gifford sealed the deal by making a strategic alliance with the Queens council members. One member who'd backed the other leading contender, Angel Rodriguez, told Gifford that he'd done so because Rodriguez had promised "license plate 1," a reference to the special license plates that council members got.

Three months later, Rodriguez was indicted for a $1.5 million extortion scheme. While I was offended by Rodriguez's betrayal of the public trust, I was also puzzled and saddened. In my dealings with him, he'd always been affable and kind. His son had even interned in my office. His foolishness stunned me. Over the ensuing years, many of my colleagues would go to jail and, aside from the question of ethics, I was always amazed at their willingness to risk their careers and freedom for money they could surely have lived without.

Now that Gifford had been elected speaker, we turned to governing. The first order of business was appointing committee chairs. Naturally, I wanted to chair the Education Committee. The UFT opposed this but Gifford told them he didn't really have a choice: not only had I played critical roles in his election as a council member and speaker, I was eminently qualified for the position as a graduate of the New York City public schools and a former professor.

I was excited about chairing the Education Committee. I felt that I could really make an impact by shedding light on issues that nobody else had dared talk about. In the two years that I'd been sitting on the Education Committee, the press had largely ignored our hearings, and for good reason. We rounded up the usual suspects, asked the usual questions, and elicited the usual answers. It was like a Greek play: you knew the ending in advance. It wasn't about investigating or fact-finding but about reaffirming faith in the Democratic educational catechism: more money, smaller class size, and teacher autonomy. As a former academic, I believed in the power of ideas and the value of robust discourse. I wanted to get people thinking, and by that I meant not just my colleagues but everybody: the administration, the press, nonprofit groups, people working in the schools, and ultimately the general public.

I also knew how lucky I'd been to get this position. Sure, I'd worked hard, but I'd also gotten many good breaks: coming across that article in the *Times* about Gifford's election just when I had time on my hands; running against an incumbent who soon resigned; supporting Gifford for speaker. I'd been fortunate to land this position and I was determined to make the most of it.

26

MY TEAM IS ON THE FLOOR

2011–2012

Replacing a principal at Success is a big deal. We have to identify someone with sufficient drive and talent, train her, and then help her get to know the school community before taking the reins. To allow time for all this, we ask principals who intend to leave to give us as much advance notice as possible, ideally two years. Thus, it came as a great surprise when, on April 25, 2011, less than two months before the end of the school year, Harlem 3's principal told me she wouldn't be returning. That wasn't enough time to train a principal even if we had somebody who was ready to be trained, which we didn't. But somebody had to run Harlem 3 so I chose Harlem 4's dean, Richard Seigler. Richard was only twenty-four years old and had just two years of teaching under his belt, one as a dance instructor, but I'd come to believe he had real leadership potential. He'd first come to my attention when I'd seen his students give a dance performance that was far better than those at our other schools. This flowed from Richard's self-confidence and sense of purpose. In Richard's first year with us, he'd noticed that students were occasionally missing his dance class to meet with other teachers about behavioral issues. Many brand-new teachers would have been too intimidated to object, but Richard wrote his colleagues:

> Dance is part of the curriculum and all my instructional time is valuable. If a child is out of control, please feel free to reprimand

them during your own classroom time. Thank you in advance for your cooperation.

Richard wasn't afraid to ruffle a few feathers and that's a critical quality in a leader. Principals who worry too much about being popular end up following rather than leading. While leaders must treat teachers fairly, they need to make their own decisions about what is fair and reasonable, not look to others to tell them.

But despite Richard's potential, I wouldn't have promoted him so quickly if I could have avoided it. Before becoming an elementary school principal, a teacher should have several years of instructional experience in at least a couple of different grades to understand the scope of the elementary school curriculum and how young children learn. We'd have to give Richard a crash course in this, so I emailed Paul Fucaloro, copying Keri Hoyt:

EVA: *Richard will need a lot of support! Hope you can help!!!!*

PAUL: *Fine! We will need to know his knowledge base in ELA and math.*

EVA: *Zilcho.*

PAUL: *Oy vey iz mir!!!!! He better be a quick learner. I can meet him on Fridays and give him specific things to study during the weekends. We can visit classes together, explore lesson planning and effective lesson delivery, then assessment/reteaching plan. He must get a feel of what Literacy and Math instruction should look like as scholars ascend the grades.*

EVA: *Would be a huge help.*

PAUL: *Done! May I work starting this Friday at Harlem 4 from 10–12? Should I start both on Literacy and Math in K and 1 this week, then give a study assignment, which I will forward to you Keri, in advance?*

KERI HOYT: *Would you mind meeting at 3? Richard's plan is to help work the door at Harlem 3 over the next couple of weeks to learn the families' names, etc.*

EVA: *I would focus on 1st grade. Frankly K and 1 are in worse shape at Harlem 3.*

Then, at 4:27 p.m. the following day, just twenty-seven hours after my first email, Paul sent a comprehensive plan for training Richard. In that brief period, we'd figured out how, where, when, and who would train Richard. I believe this nimble and non-bureaucratic approach to solving problems is a key to success.

But another thing worried me. Harlem 3 was our weakest school because the outgoing principal had placed teacher happiness ahead of student achievement. While good teacher morale is important, a school's ultimate priority must be providing students with an excellent education, and achieving excellence in any field requires some amount of sacrifice. Most of our teachers ultimately embrace this sacrifice when they come to understand how much it improves their instruction and the impact it will have on the lives of their students. However, just as many people need a personal trainer to help them achieve their fitness goals, teachers often need a principal to push them to reach their potential, to help them understand the intensity that excellence requires. Harlem 3's former principal hadn't done that because she'd never truly been on board with our philosophy. Richard would therefore have the unenviable task not only of replacing a popular principal but of informing teachers that their instruction might not be as good as they'd been led to believe.

Knowing some teachers might not want to return next year, my HR staff wanted to convince them to stay on, but I discouraged this:

> It does no one any good to have unhappy campers on the bus. We will not be setting Richard up for success if people are grumbling. If teachers or staff are hesitant they will quit [later]. That will be far worse for the kids.

Rather than pressuring teachers to stay, we told them they should return only if they were truly committed to the Harlem 3 educational community and to let us know their decision as promptly as possible so we could hire replacements if necessary. Some did choose to leave, but most decided to stay on.

That summer, we got our test scores and, as I'd feared, Harlem 3's English passage rate, 72 percent, was the weakest of our four schools. By contrast, 85 percent of Harlem 1's students had passed the ELA test. Then, as if the challenge facing Richard wasn't big enough already, teachers who'd previously told us they'd be staying began resigning. By the second day of school, nine had resigned, virtually all to work with Harlem 3's former principal, who'd taken a job at another charter school. I didn't mind their resigning, but I did mind that, despite our pleas, they were doing so at the last minute when good candidates would be hard to find and would miss our teacher training. I also felt terrible for Richard since this exodus must surely have felt to him like a vote of no confidence. Then, to add insult to injury, Richard's business operations manager resigned on the first day of school.

I worried I'd dealt Richard an impossible hand, particularly given his lack of experience. At a minimum, he'd need enormous support to succeed. To begin with, we'd have to find him a new business operations manager (BOM) to help him with logistics such as making sure forms were collected and figuring out how to move kids around the building in an efficient manner. While these things may sound mundane, they really matter. Without permission slips, students can't go on field trips; without a good plan for getting kids to and from the lunchroom, traffic jams occur that cause misbehavior, mayhem, and lost instructional time. To ensure that Harlem 3 didn't descend into chaos, Kristina Exline, who was in charge of all of the BOMs, agreed to fill in for the missing BOM while she trained a new one.

I then asked our director of literacy, Arin Lavinia, to make training Richard her number one priority. She first helped Richard master the literacy curriculum that Arin and I had designed. It had nine components including shared text, in which the whole class reads a brief passage together and then writes about it; read alouds, in which the teacher reads from a book, interspersing questions to help the students with their comprehension; and guided reading, in which a teacher helps a half dozen students who are at exactly the same level read a book that is slightly above that level.

Arin also helped Richard master the two main methods by which our principals improve instruction. The first is leading planning meetings, which are like graduate seminars. Rather than assume that our teachers have fully understood a text, we have an intellectual group discussion to help them do so. We identify the essential meaning of the text and the craft and structure the author employs. This approach reflects our belief that great teaching requires far more preparation than most educators realize.

Principals also improve instruction through coaching, which itself comes in two flavors. One is coaching in the moment: as the teacher teaches, the principal makes suggestions, such as telling the teacher to listen in on students' "talk and turns." If the principal feels that the teacher needs a demonstration of a particular teaching technique, the principal may even take over instruction for a few minutes to do so.

Principals also give feedback after teacher observations. While all schools do this, we have a much quicker feedback loop. In an email the principal drafts during or immediately after the observation, she makes concrete suggestions, such as giving students more time to think of an answer before calling on students, and then checks in on the teacher later that day or the next day to see if these suggestions are being implemented successfully. If so, the principal commends the teacher and makes additional suggestions. This continuous improvement and quick feedback cycle works wonders. In district schools, the teachers' union contract only allows teachers to be observed twice a year. Our teachers are observed daily by either a principal or assistant principal. We can do this in part because we keep the observations short, typically 5 minutes, which is enough time to come up with concrete suggestions for improvement. Principals also provide feedback based on their study of students' test results and written work.

This work—coaching teachers and leading planning sessions—is the heart of a principal's job at Success. If I find a principal sitting alone in her office every time I visit a school, I worry that she doesn't understand the true nature of her job.

Arin led planning sessions at Harlem 3, so that Richard could see how to do them, and then prepared him to lead these planning sessions himself. She also showed him how to tell the difference between good and bad teaching and how to provide effective feedback.

Richard found that many of his teachers just weren't expecting enough of students both intellectually and behaviorally. Richard had very high standards for both teachers and students, which I suspect was in part due to his background as a professional dancer. Dancers know that achieving excellence requires hard work, concentration, precision, and lots of practice. Richard saw that rather than insisting that students do things and giving consequences if they refused, teachers were pleading with students. The teachers also lacked urgency. At Success, we're always hustling because time is a finite resource. We try, for example, to minimize the amount of time between when one class ends and instruction at the next class begins. At Harlem 3, however, kids were sauntering between classes and then taking forever to settle down and start learning. Richard's impressions were confirmed by Paola Zalkind, a principal in training at Harlem 3, who saw that it lacked the intensity and commitment to excellence of the two other Success schools at which she'd taught.

I visited Harlem 3 frequently to see how Richard was doing and I often saw the father of one of our students sitting in the lobby. One day I spoke to this man, Emile Yoanson, and he explained to me that, due to his poor health, he didn't want to travel to the school twice from the Bronx, where he lived, so he just spent the whole day sitting there. This shows the incredible lengths to which many of our parents go to provide their children with a good education. I further learned that although this man was married, his wife and another son were in the Ivory Coast where a civil war was raging. Given my own family's background, I very much empathized with their plight, so I reached out to Senator Chuck Schumer, who helped us get visas for them. Then it turned out they didn't have enough money for airfare, so we paid for that as well.

On December 6, they arrived at Kennedy airport. A week later, Mr. Yoanson's younger son, Segnonble, who didn't speak a word of English, joined his brother, Christian, at Harlem 3. The family was overjoyed to be reunited after so many years and thanked us profusely. When all of the opposition to our schools gets me down, it's little moments like these that keep me going.

Meanwhile, Richard worked relentlessly to improve Harlem 3. Many students were arriving late, which caused them to miss instruction and distracted other students. To discourage lateness, Richard would regularly man the entrance where students arrived and confront parents who were bringing their children to school late. Some of these parents began dropping their children off at the corner so Richard responded by going to the local subway station to confront them. This very much reflected our "by any means necessary" approach to problems.

Unfortunately, some teachers resisted Richard's efforts to improve student behavior. Even though Richard is African American, they claimed that Richard was reinforcing negative stereotypes of minority kids by demanding that they speak respectfully. In fact, Richard learned from some faculty members who supported him that other faculty members who didn't had decided to make his life as difficult as possible so he'd resign. They did so by acting disrespectfully toward him at faculty meetings, implying he lacked the qualifications to lead the school, and refusing to follow his directions. While it was undoubtedly true that Richard was hindered by his inexperience, undermining him didn't help, particularly since I didn't have a more experienced principal waiting in the wings to take over if Richard quit.

Richard could have decreased opposition to his leadership by being more deferential to his teachers and allowing them to teach the way they wanted, but he knew this wasn't why I'd made him a principal. Fortunately, he was courageous and determined not to let adult politics get in the way of giving his kids a great education. Some teachers, however, just weren't willing to play ball with Richard, which presented him with a real dilemma. The last thing Richard needed was to lose teachers midyear but having teachers who refuse to follow the playbook is a big problem even if the teachers are competent. There is a scene in the wonderful basketball film *Hoosiers*. The best player on the team is making baskets but he isn't following the team's strategy of passing the ball before shooting, so the coach benches him. When another player fouls out, the team is left with only four players on the floor. "Coach, you need one more," says the referee, to which the coach replies, "My team is on the floor." I feel the same way. No matter how good a teacher is, if that teacher won't play as part of the team, you're better off without her. Mind you, it's not a question of who is right. Reasonable minds can differ as to the best approach to pedagogy. However, a school's educators need to be working from the same playbook.

It's easier to achieve this at a charter school. In district schools, wars will break out between competing pedagogical philosophies. With a charter school system, however, teachers can choose a charter school that reflects their philosophy. This, indeed, is exactly what had happened with Harlem 3's former principal and the teachers who had left with her. But there were apparently still some teachers at Harlem 3 who weren't on board with our philosophy and wouldn't follow Richard's direction, so he let them go with my full support.

From this point on, it was a tremendous amount of work but at least Harlem 3's teachers were all pulling in the same direction and things gradually improved. That summer, we got our test results. Harlem 3's English passage rate jumped from 72 percent to 89 percent and its math passage rate held steady at 93 percent. This incredible outcome was a testament to Richard's determination and abilities given the turmoil the school had suffered, including the loss of dozens of staff members and the opposition of many who remained. It was also a testament to the support that Arin, Paola, and Kristina gave Richard. Three years later, Harlem 3 won the prestigious National Blue Ribbon Award.

Interestingly, there was also a big improvement in the scores at the charter school that had gotten all of our teachers. Its English passage rate went from 33 percent to 77 percent in third grade and from 30 percent to 58 percent in fourth grade. I was glad that children at that school had benefited from the training we'd given their teachers, I just wished it had happened under different circumstances.

Our year was marred by a tragic event. Christian Yoanson had been hospitalized with an illness and on June 8, we learned he'd died. Jenny Sedlis and I went to visit Christian's parents at their home to console them. When we arrived, Mr. Yoanson had just returned from signing Christian's death certificate and was unable to speak. I held him in my arms while he sobbed uncontrollably. When he stopped, he and Mrs. Yoanson talked about how much Success had meant to Christian. Right up until the end, they said, Christian had been talking about his eagerness to return to school. The walls, I noticed, bore not only many examples of Christian's schoolwork but framed copies of newsletters and other materials we'd sent home.

Christian's family was completely overwhelmed by this loss and our entire community pulled together to help them through this difficult time. I asked my special assistant Maria Monforte Trivedi to arrange the funeral and help find a cemetery plot. Since cemetery plots in New York City were hard to come by, Maria's own family donated an extra plot they owned. If you visit that cemetery today, you will find buried amid the Monfortes, a family of Italian descent, the grave of Christian Yoanson, an immigrant from the Ivory Coast taken at a tragically young age.

While witnessing the Yoansons's pain was heartbreaking, I took a small measure of comfort in knowing we'd at least reunited Mr. Yoanson with his wife and younger son, Segnonble, who, as I write this now, is a sixth-grader at Success whose accomplishments reflect his family's great commitment to education.

27

GIVING AN HONEST DEFINITION TO THE WORD "PUBLIC"

2011–2012

I have to admit it: I've got a soft spot for Matt Damon. *Good Will Hunting, Ocean's Eleven,* and those Bourne films . . . *Sigh.* But, alas, Damon doesn't have a soft spot for me. At a rally in DC in the summer of 2011, he criticized me by name. It's a strange experience watching a guy you're used to rooting for in the movies make clear he isn't rooting for you, but at least I was in good company. Obama had lamented: "I love Matt Damon, love the guy. Matt Damon said he was disappointed in my performance. Well, Matt, I just saw *The Adjustment Bureau* [Damon's latest film], so right back at you, buddy." In my case, Damon didn't like the fact that we sent mailings to parents about our schools. I confess I found it curious he would begrudge my spending money telling disadvantaged families about their educational opportunities when far greater funds were spent on promoting the films in which he appeared. (Right back at you, buddy.)

Notwithstanding Damon's opposition, we decided in the fall of 2011 to open three new schools in Brooklyn. Many people warned me that continuing to expand so quickly could hurt the quality of our schools. I was conscious of that danger, but also of the fact that there were more than one million school-age children in New York City who desperately needed better educational options. We'd received nine thousand applicants for just nine hundred seats. We had a moral responsibility, I believed, to haul as many children into our lifeboat as we could without sinking it.

One obstacle we faced was raising enough money to fund this rapid expansion, but a strong prospect soon emerged: a blunt and brilliant investor by the name of Daniel Loeb. He'd started his business career at twelve, making skateboards and investing the profits in the stock market, earning him the sobriquet Milo Minderbinder. By the end of college, he'd amassed $120,000, which he then proceeded to lose in one disastrous trade. Nonetheless, he went into finance and, in 1995, founded a hedge fund. Two years later, it earned an astonishing 98.3 percent annual return. By the time I met Daniel, he was managing billions of dollars.

Daniel and his wife, Margaret, were also deeply committed to the causes in which they believed. They had played a major role in passing gay marriage legislation and were generous supporters of Prep for Prep, an organization for which I'd worked that helped gifted minority children. Daniel was looking around for a charter school to support and was going about it with the same thoroughness with which he chose his investments. He eventually concluded that Success Academy was the equivalent of an undervalued company: we were getting great results for kids, but we weren't as popular with some funders as other charter schools because I was controversial. Since Daniel and Margaret cared about helping kids more than winning popularity contests, they generously agreed to give us $3 million, enough money to open all three of our Brooklyn schools. No Success funder had ever done that before.

One of our schools, we decided, should be another mixed-income school and Jenny found the perfect location: Cobble Hill in District 15. This district had become increasingly diverse as more affluent families had come to the neighborhood, but most of its schools didn't reflect this. While gentrification brought improvements to a few schools, it was quite uneven, and rising home prices often resulted in poor residents being pushed out of the neighborhoods zoned for these schools just when they improved. Enrollment of poor children in the two best schools in District 15 had fallen dramatically since 2004: from 62 percent to 17 percent at PS 58 and from 36 percent to 19 percent at PS 29. Poor children were generally relegated to schools like PS 15, where 95 percent of the students were poor and student achievement was dismal. Unfortunately, this tendency toward increased segregation reflects a national trend. From 2001 to 2014, highly segregated schools—those with more than 90 percent low-income students and students of color—doubled in number.[23]

Given the considerable racial and economic segregation in District 15, we wanted to create a school that would serve a cross section of the district's population, from residents of public housing to owners of multimillion-dollar brownstones. We also decided to open a school in Williamsburg, another diverse community in northern Brooklyn.

On October 29, we held an information session regarding one of our proposed schools. Dozens of parents showed up, including one woman who told the *Daily News*: "I'm not zoned for the best school, so . . . one more . . . good educational option in this neighborhood is a very positive thing." When I started my presentation, however, some people in the audience started heckling me. "If you can't hear me out," I said "we'll have to cancel the meeting because I can't shout over you." But the heckling continued so I finally had to give up. One parent said afterward, "I'm just here to learn. It's a joke. We're not allowed to go to a forum to learn about something? We have a right to understand our options."

Another woman named Ismene Speliotis, however, said she was "happy with the way Eva left." This woman, who'd organized the protest, described herself to the press as simply a parent and PTA member, neglecting to mention that she was also the executive director of ACORN's housing affiliate.

The usual misinformation campaign followed. At a December 14 hearing, a teacher at the school where we would be co-located claimed that the "gym was so crowded that high school students couldn't take the required physical education classes in time to graduate." This was a complete fiction. The building was massively underutilized, at only

57 percent capacity. Then, on January 11, 2012, we came across the following note on the Internet:

> January 13 at 5:30 p.m. there is an IMPORTANT MEETING for District 15 parents interested in being plaintiffs in a lawsuit against the placement of the Cobble Hill Success Academy.

Curiously, the notice didn't say who was sponsoring the meeting but the address for it was the same as that of NYCC (a successor to ACORN after it closed due to a series of scandals), and the phone number listed belonged to an employee of the Alliance for Quality Education (AQE), one of the UFT's many front groups.

We asked a District 15 parent we knew to attend and while he was quickly kicked out, he learned the meeting would be led by a woman named Megan Hester who did community organizing for the Annenberg Institute, an affiliate of Brown University. It turned out she was one of five Annenberg employees, two of whom were previously associated with ACORN, to "provide research, training, and logistical support to the New York City Coalition for Educational Justice [CEJ]." CEJ was yet another member of the alphabet soup of union front groups. It fought virtually all of Mayor Bloomberg's education reform policies and "organiz[ed] parents in schools . . . slated for closure or co-location." Annenberg claimed that its role was merely to provide "strategic and technical support," but, in reality, it ran CEJ. While CEJ's Web site claimed it was "led by parents," there wasn't a single parent listed on the site and its "members" were a collection of unions and antireform interest groups including the UFT, NYCC, and The New York Civic Participation Project (which was itself a "collaboration of labor unions and community groups organizing union members"). CEJ's meetings were held at Annenberg's New York City office, and their phone numbers were almost identical. Finally, although CEJ's reports didn't list their authors, the meta-text shows these reports were in fact written by Ms. Hester and her colleagues at Annenberg. In short, Annenberg didn't help parents express their views, it just passed off its employees' own views as those of parents. It was their own think-tanky version of the UFT's rent-a-mob tactic.

As a former professor, I was surprised to see that a university-sponsored institute would instigate a lawsuit to prevent a school from opening. It struck me as a rather brass-knuckles tactic for an institution of higher education. It was also a betrayal of Annenberg's founder, the late Dr. Theodore Sizer, a well-respected educator. I'd served on a panel with Sizer, and he'd expressed enthusiastic support for charters and what Success was trying to accomplish. Indeed, he'd founded a charter school himself and supported the concept of a "market" in education. "Parental choice among schools," he said, could play a critical role in advancing school quality since "[f]amily interest represents an important kind of assessment." He'd also written that a charter school "open on a lottery basis to any child wherever that child may live—rich suburb or poor city—finally gives an honest definition to the word 'public.'" This was exactly what we were trying to do in Cobble Hill—to create a diverse school by drawing students from a broader geography than most district schools. But now the institute Sizer had founded was trying to stop us from fulfilling his vision.

The mastermind behind Annenberg's anti-charter organizing was Norm Fruchter, a former editor of the *New Left Review* who had defended Fidel Castro's refusal to hold elections on the grounds that if people didn't like it, they could just have another violent revolution. Fruchter's role at Annenberg would have particularly appalled the late Walter Annenberg, a conservative businessman who'd donated $50 million to the institute named after him. I believe it would also have appalled Bill Gates and the late Harry Helmsley, businessmen whose foundations all supported charter schools but had nonetheless contributed to the Annenberg Institute, no doubt unaware of Annenberg's clandestine anti-charter activities.

Annenberg's lawsuit against Success was filed on February 8 by Arthur Schwartz, ACORN's former general counsel, who claimed we shouldn't be able to start a school in District 15 because we'd told SUNY that we expected to open our school in a neighboring district before finding the Cobble Hill building. While we'd obtained SUNY's approval to change districts and gone through the city's whole process of notices, hearings, and approval to get co-located, this wasn't enough red tape for Schwartz.

Three weeks later, Schwartz brought a lawsuit against our Williamsburg school, this time claiming we hadn't done "meaningful community outreach." The lead plaintiff, Council Member Diana Reyna, claimed that our "marketing strategies . . . discriminated against Latino and low-income families." Another opponent claimed our marketing had been "racist" and that we "only want[ed] the white middle-class" children in our schools.

Then, the *Times* published an op-ed piece titled "How Charter Schools Can Hurt" by Lucinda Rosenfeld, the parent of a student at PS 261, who claimed that if Success Academy Cobble Hill attracted any middle-class students, it would destroy PS 261 and "create a snowball effect in which [the] middle-class population ends up fleeing" the city. This middle-class flight Armageddon scenario struck me as far-fetched given the pace of gentrification in Brooklyn. Surely there were enough middle-class families to populate one more diverse school, and other families in the district weren't as satisfied as Ms. Rosenfeld with their existing options.

While I tried to fend off this public relations assault, Emily Kim and her team drafted our legal papers. Luckily, she was able to reunite the old band—Paul, Weiss and Arnold & Porter—and we'd carefully documented our extensive outreach efforts. We'd learned from bitter experience that politicians would ignore our phone calls and then claim we'd never reached out to them, so we'd made sure to put everything in writing this time around. In the case of our Williamsburg school, we'd written letters to twenty-six elected officials and community leaders including Council Member Reyna, the lead plaintiff in the Williamsburg lawsuit. In addition, we'd collected 8,300 signatures; repeatedly appeared before both the Community Education Council and the Education and Youth Services Committee of the local community board; mailed out numerous brochures; purchased outdoor advertisements on bus shelters, phone kiosks, and subway stations; put flyers under doors in public housing projects; and held eleven information sessions and twelve school tours. We'd also sent out five brochures in Spanish and advertised in Spanish. We explained all this in our legal papers, which we submitted in the Cobble Hill litigation on March 9 and in the Williamsburg litigation on March 23.

Emily also recruited parents whose children had applied to our schools to intervene in the lawsuit. Schwartz opposed the intervention: he was all for parents having a voice—unless they supported charter schools.

In the reply brief he submitted on April 2, Schwartz argued that it was our fault that his clients had ignored the letters we'd sent them because we'd sent "only one letter to each elected official," rather than pestering them with multiple letters and hadn't warned them "that if they fail[ed] to respond, they m[ight] not have another opportunity," as if they were children who needed reminders to do their homework.

Meanwhile, despite claims that the community didn't want our schools, applications poured in. Demand was particularly strong for our Cobble Hill school, which received 1,457 applications for just 172 spots. (Success Academy Upper West, by the way, now had 2,186 applicants for 89 seats, a higher ratio than Harvard.)

On May 30 and June 1, the court ruled in our favor in both lawsuits, finding that our outreach had been sufficient and that there was nothing untoward in our having switched Cobble Hill to District 15. But we faced one more obstacle before we could open. The law now required that if we renovated a facility, DOE had to spend an equal amount on the co-located school. It sounded fair, but in reality it led to wasteful expenditures. For example, at Upper West, we'd turned some classrooms into a lunchroom since we needed one. DOE therefore had to spend money renovating our co-located schools even though they didn't need to build a lunchroom or have any other compelling needs for renovations. Now we needed to make renovations at our Brooklyn schools including electrical upgrades to run our SMART Boards, laptops, and air conditioners (since our school started in mid-August, which could be brutally hot). DOE, however, refused to let us make renovations we *did* need so that it wouldn't have to spend money on renovations our co-located schools *didn't* need. Moreover, it wasn't just us; all charter schools were being prohibited from renovating their facilities.

I reached out to Mayor Bloomberg and he called me back—right in the middle of my nephew's bar mitzvah. While I feared my family would never forgive me, I didn't know whether I'd get another chance to talk to him so I stepped out and took his call. Bloomberg said the city just didn't have the money to match the renovations that charter schools wanted to make. I responded that having served on the council, I knew full well that the city had an annual capital budget for schools of $2 billion and that the amounts at issue were trivial in comparison. If he didn't let us make these renovations, I said, I'd install emergency generators and call the press to let them take pictures. The next sound I heard was a click of the phone hanging up.

About a week later, I got word that the administration had changed its mind: all charter school renovations, including Success's, were now being approved. I don't believe, however, that I had forced the mayor's hand. Bloomberg wasn't someone who could be pushed around. Rather, I believe my threat to make a fuss simply caused everyone to take a harder look at the situation and they concluded that allowing us to make these renovations made sense. After all, even if there was some waste of public funds, the net result would be a lot of renovations to city-owned buildings, much of which charter schools would pay for.

Sometimes, government needs shock therapy because lower-level bureaucrats don't get the big picture. Around this same time, the US Department of Education suddenly decided that we were violating federal law by giving a preference in our lottery to English language learners (ELLs). I didn't believe that Secretary Duncan himself would actually support this inane policy, so I wrote him:

> The Department is threatening to withdraw millions of dollars. . . . This unconscionably bad policy imposes a cruel Sophie's

> choice on Success. Either we must abandon our ELL preference, and serve fewer students from this needy population, or we must cut millions of dollars from our program, which will harm most of all these very same needy students. [This] threat[] is a gun pointed at our head.

I used such strong rhetoric so I'd be sure to get Duncan's attention. The policy was soon changed.

In late August, our new Brooklyn schools opened. We were able to give the district another sorely needed integrated school. In 2016, our Cobble Hill school had precisely 50 percent free and reduced-price lunch students and very much reflected the district's diversity with 27 percent white students, 26 percent Hispanic students, and 38 percent African American students. As for our Williamsburg school, despite the claims that we were only trying to serve white middle-class families, only 9 percent of the students are white and 75 percent of the students are eligible for free or reduced-price lunch.

Remember Ms. Rosenfeld's claim that we'd destroy PS 261 and cause middle-class flight? Well, surprise, surprise, it didn't happen. As of 2016, PS 261 had exactly the same percentage of free and reduced-price lunch students it did the year before our school opened (44 percent). Indeed, PS 261's principal commented that our school had "not had a negative" effect on PS 261 and that it was "really nice that parents have the ability to choose what they believe is best for their children and family, and what complements their belief about education." Finally, we didn't cause middle-class flight and turn Ms. Rosenfeld's neighborhood into a slum. By 2016, town houses in her neighborhood were going for $4 million a pop.

Many people have accused Success of having terrible motives and have predicted that all sorts of horrible things would happen when our schools opened. These accusations and predictions have never proved accurate, but we have yet to receive an apology from the people who made them. I'm not holding my breath.

28

COMPETING IN THE NATIONALS

2011–2012

If you want to understand the level of our teachers' commitment, witness the tale of Lydia Cuomo, a woman from New Hampshire whom we hired to teach at one of our Bronx schools. On August 19, 2011, she failed to show up for the first day of school and we soon learned why. Just blocks from the school, she'd been waylaid by a drunken man who had forced her into an alley, pulled out a gun, told her he'd "blow [her] head off" if she didn't do what he said, and raped her. As he attacked her, Lydia saw something unimaginable: a badge. He was, it turned out, a New York City police officer and the gun he was holding to her head was his city-issued firearm.

It's hard for any woman who's been raped to feel safe again but imagine how much harder it is when the rapist is a policeman, someone whose presence on a dimly lit street is supposed to make her feel safe. One could have hardly blamed Lydia if she'd packed up her bags and returned to New Hampshire or looked for a job in a safer neighborhood, but she did neither. Rather, just weeks after the assault, she began teaching at our Bronx school. She also testified at a trial against her assailant and publicly advocated for better legal protections for rape victims. When she did so, she used her real name, an act of personal bravery and a public statement that the only person who need feel shame when a rape occurs is its perpetrator.

Sometimes teaching in a poor neighborhood can be risky and some of our educators have even been attacked by parents of the children they serve. A security camera caught an assault on Harlem 1's principal and it was horrible to watch. The principal lay on the floor futilely trying to fend off the blows of a parent about three times her size until one of our staff members managed to pry the parent off her. The emotional impact of such attacks on educators can be highly damaging but educators can't afford to shy away from difficult parent conversations for fear of being assaulted.

While we struggled with these issues, we also tried to improve our school design. At the beginning of the 2011–2012 academic year, I asked Sean O'Hanlon, our chess czar, to make our chess program the best in the country and he set about doing so. With 3,500 scholars in nine schools, some of our students were ready for stiffer competition so he decided that this year we should compete in the national chess tournament that took place in Nashville, Tennessee. This worried me since most of our opponents would be from suburban and private schools. I feared we'd get a drubbing that would undermine our students' confidence and pride in their abilities. Sean assured me that he would improve our program to increase our scholars' chances. He began running tournaments on weekends so our top players at different schools could compete against one another. He hired better chess coaches so our students could get the more advanced instruction they needed as their play improved. He switched his chess staff to a Tuesday-through-Saturday schedule so students could practice on the weekend.

Another area in which we made improvements was in our selection of books. I'd first realized their inadequacy when Dillon became a student at Success. We'd created a summer reading list with hundreds of suggestions, but when I read them to Dillon, I found that many were mediocre. The quality of books matters enormously. When teachers read certain books aloud, such as *Charlotte's Web* or *My Father's Dragon*, the students hung on every word and would beg the teacher not to stop. Great children's books are also more instructive. Take Dr. Seuss's books, which are like Shakespeare for children. Their rhymes and rhythms help children appreciate the beauty of language and understand how it works. Consider this stanza from *The Cat in the Hat*:

I know it is wet
And the sun is not sunny.
But we can have
Lots of good fun that is funny!

Dr. Seuss is implicitly teaching children that words can have noun and adjectival forms (sun and sunny). Moreover, as a child rereads this book and becomes more sophisticated, he may see Dr. Seuss's subtle play on words: that unlike "sun" and "sunny," the words "fun" and "funny" don't have the same meaning.

Thus, just like adult books, children's books can be great and we are fortunate to be living in a golden age for children's literature. We have classics like *Where the Wild Things Are, Stuart Little, James and the Giant Peach, Where the Sidewalk Ends,* and *A Wrinkle in Time,* and more recent additions like *The Giver, The Book Thief, Monster: The Autobiography of an L.A. Gang Member,* and *The Absolutely True Diary of a Part-Time Indian.* Not all books are created equal and children should be encouraged as much as possible to read the best ones.

I found books for my own children at the Bank Street Book Store where a woman named Sara Yu always had great suggestions. I asked her to pull together a list of books for Success and it was so good I hired her to be our director of children's literature. Here is what Sara wrote in one of our newsletters about her job:

> My colleagues and I spend hours and hours discussing and selecting the best books to use for a particular unit. Earlier this year, I read more than 500 poems in one day to find the 20 or so that our middle school scholars would study that month. We consider lots of factors: the quality of the writing, plot and character development, accuracy in nonfiction, the Common Core standards, the goals of that particular unit, etc. But the final verdict usually sounds something like: "This is such a fifth-grade book—the kids are going to really get Kenny," "Second-graders will think Clementine is hilarious," or "Nope, I just don't think kids will love it."

Despite Sara's talents, it's unlikely a district school would have hired her because she didn't even have a college degree much less formal teaching credentials, just talent and a love of children's literature. Charter schools, however, have more flexibility in hiring.

Another area in which we sought to make improvements was test preparation. Many students dread standardized tests, which is unfortunate. Tests are a lot like an athletic competition: you practice hard, prepare yourself mentally for game day, and then you get to show off your skills. Since schools hold pep rallies to get everyone excited about big athletic events, I decided to hold a "Slam the Exam" rally at which our students exuberantly chanted and sang songs like Michael Jackson's "Thriller" and Queen's "We Will Rock You." It instilled enthusiasm and confidence in our students.

But paying all of the people who were making these improvements cost a lot and we found we were running a deficit at the Network, the nonprofit entity which performed functions that it made sense to centralize, like designing the curriculum, training teachers, and managing renovations and technology. While each school was chipping in 10 percent of its budgets to cover the Network's cost, it wasn't enough. Unless we cut back on these services, which we felt would do great harm to our schools, we needed to increase the Network fee. We sought SUNY's permission to increase this fee to 15 percent, which prompted Juan González to write that:

> SUNY trustees are rushing to approve a whopping 50 percent increase in the annual per-pupil management fee the state pays to one of the city's biggest and most controversial charter school operators.

In fact, we weren't seeking to increase what the "state pays," just change how that money was divvied up between our schools and our Network. We sought a correction but González responded, "My column did not say you were seeking an increase in state aid." What about his statement that we were seeking an "increase in the annual per-pupil management fee the state pays"? That was accurate, he claimed, because the money for the fees the school paid ultimately came from the state. It was like saying that if a congressman bought a Mercedes with his government salary, it would be fair to write "government pays for congressman's Mercedes." González knew full well that readers would think the government was paying Success more in total. In fact, after González's column, a prominent education activist wrote that we were trying "to reap more fees from taxpayers." All this spooked SUNY, so they put off the vote on our much-needed fee increase.

While this battle played out, thirteen of our kids flew to Nashville to compete in the national chess tournament, which had 1,500 competitors in total. I went to see it and was astonished at our students' powers of concentration. They played three games every day and each game could last up to four hours. The key to success, Sean told me, was patience: carefully analyzing each move and refraining from moves that might seem good at first blush but which had a hidden flaw such as leaving a piece unprotected. Playing well for up to twelve hours a day requires incredible powers of concentration and endurance, which is why chess is such valuable intellectual training.

Sean had selected Harlem 2 and Harlem 3, our two strongest teams, to compete in the tournament. By the tournament's second day, Harlem 2 was in second place, trailing the top school by only a half point, and Harlem 3 was in third. By the third day, Harlem 2 had jumped into first place and Harlem 3 had fallen to fourth. In the final round, Harlem 3's strongest player, my son Dillon, was matched up against Harlem 2's top scholar, Jameek. Dillon won, which unfortunately, knocked Harlem 2 out of first place, but it still placed second, and Harlem 3 placed fourth. Individually, Dillon placed second, winning seven out of eight games. As both a parent and a school leader, I was proud. Here we'd never before competed in the nationals and our schools had finished second and fourth!

I soon learned that our students had also done well on the state tests that year. Eighty-eight percent had passed in English and 97 percent in math. This was more than twice as well as the district elementary schools in Central Harlem, in which 32 percent of students passed in English and 40 percent in math.

As for our fee increase, we'd gotten SUNY to put it back on the agenda for its June 25 meeting. González again opposed it, claiming it would be "the rich . . . getting richer," which made it sound like we'd be spending the money on caviar and champagne, not providing a better education for disadvantaged kids, and wasn't even true since the total amount of money wouldn't increase. Fortunately, SUNY ignored González this time around.

González's opposition was astonishingly cynical, even by his standards. He knew full well that our fee increase wouldn't cost taxpayers a dime nor did he ever identify any other harm it would cause. Thus, I can only conclude that he opposed our fee increase solely because he wanted to hobble us in any way he could, even if doing so would harm children. Over the years, González's monomaniacal hatred of Success led him to write twenty-four negative columns about us. González is both smart and industrious and could have rendered an enormous service to New York City by trying to shed light on why hundreds of its schools were failing to provide their students with an adequate education. Instead, he sought to tear down a network of schools whose students were mastering rigorous academic content, acing state tests, winning national chess competitions, and whose teachers were making great personal sacrifices, even putting their safety at risk, to help students attain these achievements. What a sad waste of his talents.

29

THE REDHEAD WITH THE VORACIOUS APPETITE FOR DATA

2002–2004

At noon on January 1, 2002, I sat in the audience as Mike Bloomberg, New York's billionaire businessman, gave his inaugural address on the steps of city hall. Four months earlier, the twin towers would have been clearly visible from where Bloomberg stood; now they were just a heap of rubble being carted off by a never-ending procession of dump trucks. But while these towers didn't loom over Bloomberg's inauguration physically, they did psychologically. September 11 had both injured the city's economy and blown an enormous hole in its budget. Belt tightening, warned Bloomberg, would be necessary, but "even though we must sacrifice now," he added, "let us not forget we are still a city of big dreams, of big ideas, big projects, and a big heart."

Bloomberg also indicated that he intended to obtain greater control over the city's school system. The existing governance structure had come about in response to community demands for greater local control culminating in December 1966, when protesters took over the meeting hall of the board of education and declared themselves the people's board of education. Mayor Lindsay responded to these demands with an experiment in which the residents of the Ocean Hill–Brownsville area of Brooklyn were allowed to elect a local school board to run their schools. The board they elected appointed a superintendent who soon dismissed thirteen teachers and six administrators. Terrified at what might happen if this model was adopted more broadly, the teachers' union staged a citywide strike demanding reinstatement of the dismissed teachers.

The state education commissioner resolved this crisis by taking over the Ocean Hill–Brownsville schools and reinstating the teachers who'd been dismissed, while the legislature passed a law to address the demands for more local control. Their solution was to give the borough presidents the power to appoint a majority of the members of the board of education, which would share power with thirty community school districts. The problem with this governance structure was that power was so dissipated that the buck stopped nowhere.

For decades, mayors had been trying to wrest back control of the schools but the very same features that made this Rube Goldberg structure ineffective also made it difficult to dislodge. Unions liked it because the dissipation of authority diminished the risk that any dramatic and unwelcome changes would be made. Politicians liked it because the community school boards provided opportunities for political patronage.

Bloomberg, however, was determined to succeed where his predecessors had failed and, in a speech to the council, reiterated his intention to get control over the city's schools. I supported the changes Bloomberg sought, but I didn't have a vote on the matter as the governance structure was dictated by state law. To get it changed, Bloomberg would have to deal with Albany's Wizard of Oz, Assembly Speaker Sheldon Silver, who was the apotheosis of George Orwell's dark vision of power as an end in itself. Because he was not burdened with any discernible philosophy of government or particular legislative ambitions, he could evaluate every decision with one single consideration in mind: retaining power. This power enabled him to earn millions of dollars in what were later proved to be kickbacks.

Since Silver wouldn't want to antagonize the teachers' union, Bloomberg had his work cut out for him. Moreover, when an issue was in play, Silver didn't like his colleagues to discuss it publicly since they might commit themselves to a position or, even worse, arrive at a consensus. Thus, a shroud of silence descended over Albany. I felt, however, that an important issue like this should be subject to public debate, so I arranged to have hearings on it myself even though the council lacked the legal authority to change this law. To have the desired impact, however, we'd need to have strong witnesses. The easy catches were politicians, who always welcomed the limelight, and former DOE chancellors, who were eager to share the frustrations they'd experienced with the city's cockamamy system of school governance. There were two big fish, though, that I particularly wanted to land: former mayor Ed Koch, who was well respected and would draw media attention, and Bloomberg himself, who was the protagonist in this matter. Koch readily agreed to testify. Bloomberg's advisors, however, were cool to the idea of his testifying. Just as presidents don't testify before Congress because it would make them look like supplicants before what should be an equal branch of government, mayors typically don't testify before the council. The only exception was David Dinkins, who had appeared before the council to argue for sanctions against South Africa. I argued that Bloomberg should make another exception because this would be his chance to make his case to the public. Bloomberg's aides, however, were reluctant to put a newly minted mayor in a forum over which he had little control, and in which he might be subjected to hostile questioning.

While trying to convince Bloomberg to testify, I was also struggling with how to ensure the committee examined witnesses intelligently. At the committee's last high-profile hearing, the press had commented that the committee had proved itself "worthless" since most of its members had "daydreamed" through the hearings and, when they did speak, merely "revealed how little they kn[e]w." I pondered how to elevate the level of questioning. I then remembered that in the Iran-Contra hearings, attorneys working for the committee had done most of the questioning. Moreover, it just so happened that Gifford had recently hired a Hofstra Law School professor named Eric Lane who had played a leading role in rewriting the city's charter to give the council more power. Who better to examine witnesses about governance of the school system than a lawyer who'd just rewritten the city's governance structure? Professor Lane readily agreed to play this role and then I got another piece of good news: Bloomberg had decided to testify after all.

Koch was our leadoff witness. Here was a man who'd run the city when I was a kid, a larger-than-life personality, and he was now testifying before a committee I chaired. He began his testimony by noting that as a result of "term limits and the new people, there is a spirit which didn't exist here for the last fifteen or more years." This was gratifying to hear given how hard Gifford, Chris, and I had worked to improve the council. Koch then proceeded to testify that the current system of school governance was "shameful" and that "the chancellor should be no different than a commissioner of any other agency," meaning that he should be directly appointed by the mayor. Following our plan, I let Professor Lane ask Koch questions, didn't ask any myself, and hoped that my colleagues would follow suit, which most did.

Virtually all of the dozens of witnesses who testified favored changing the law in some way, but many of their proposals were impractical. Borough president Adolfo Carrión proposed that the mayor select the board of education from individuals nominated by a civilian panel. And who would select the civilian panel? The council, the mayor, and the borough presidents, he said. Thus, he took what could be a one-step process—appointing a chancellor—and turned it into a four-step process: 1) politicians would appoint the civilian panel; 2) the panel would make nominations; 3) the mayor would select the board from these nominees; 4) the board would select a chancellor. It reflected this misguided view that, like laundering dirty money, you could take politics out of democracy by creating layers of bureaucracy. In reality, doing so just leads to paralysis and lack of accountability, which was exactly what was wrong with the existing system.

Randi Weingarten also testified, and I was curious as to what tone she'd take since she'd fervently opposed my appointment. She began as follows:

> We have had staff at your hearings . . . all week long, and I do not think in the history of the City Council [have] there been . . . hearings that are so comprehensive, that touch all bases, that really attempt to . . . be a fact-finding body. . . . It was a real public service, and I think our staff was particularly impressed by the way you did that.

While I'd like to think those comments were sincere, I suspect she was buttering me up to make amends for having opposed my appointment. On the merits of the issue, she said she supported change but was vague on the details. She was negotiating a new contract for her members so she wanted to trade her support for a generous pay raise, not give it away for free.

Bloomberg was the final witness and he argued forcefully for "abolish[ing] the board of education" and giving the mayor "sole control over the appointment of the schools chancellor, with the chancellor reporting directly to the mayor."

The hearings were widely covered and I followed up by trying to get my colleagues to publicly support mayoral control. While they were leery of doing so, especially Albert Vann, who'd risen to prominence as a proponent of local control during the Ocean Hill–Brownsville dispute, I got them to support letting the mayor appoint six out of eleven members of a board of education, which would effectively give him control, albeit with some additional transparency since the board would have public meetings. On March 20, we announced this proposal, which Bloomberg called "encouraging."

Predictably, Silver resisted any changes, and his negotiations with Bloomberg devolved into public sparring. On June 6, however, they reached a deal that was quite similar to the one we'd recommended months earlier: a board of education with thirteen members, eight of whom would be appointed by the mayor, would run the school system. It was the biggest change in the city's public school system in decades and I was excited to have played a role in bringing it about.

Soon thereafter, Eric left the law firm at which he'd been working and hung up his own shingle as he'd never enjoyed the paper pushing involved in big firm practice and wanted to get into court more. A few weeks later, he ran into an old friend of his, Bryan Lawrence, who'd recently made partner at the investment banking firm Lazard Frères and was looking for a philanthropic cause to support. They came up with the idea of starting an all-girls charter school so that parents wouldn't be deprived of that option just because they were unable to afford private school. Playing off the name of one of the most prominent private girls' schools, Eric dubbed their project "Spence for cents." Together, they recruited a board including many people from the business world and the headmistress of Brearley, a private girls' school. Their application was approved and the school opened a couple years later.

A month after Bloomberg got control of the school system, he announced his pick for chancellor: Joel Klein, a lawyer who'd led the Justice Department's Antitrust Division. Some people criticized Bloomberg for appointing a noneducator, but I felt that since the public schools were a big monopoly with all of its typical ills—lack of innovation, unresponsiveness to consumers, and a mediocre product—we needed a trustbuster. Indeed, when I first met with Klein, I told him he should think of himself as having been appointed premier of the former Soviet Union.

I felt the conditions were ripe for real change since we had a mayor who had real control of the schools and who'd appointed a strong chancellor. Moreover, as chair of the Education Committee, I could help them fix the schools by exposing problems and proposing solutions. Indeed, I soon came up with an idea that would become my single most important contribution in my time on the council.

Even though the city had built new schools with 45,000 new seats in the last decade, I'd been hearing complaints that many schools were overcrowded and I wanted to understand how this could be. Given my academic background, I loved solving mysteries and studying documents so this was right up my alley. I learned the city published something called the "Blue Book," which showed every public school building's capacity and current enrollment. I examined it and found that while many school buildings were indeed overcrowded, many others were half empty. Altogether, the city's underutilized buildings could hold 63,000 additional students, more than Boston's entire public school population. We were spending hundreds of millions of dollars to build new schools when we already had considerable excess space on our hands. I therefore wrote a *Daily News* op-ed proposing that the city "make its excess space available to charter schools to draw students from overutilized to underutilized school districts." The *New York Sun* editorialized in favor of this proposal since it would "introduc[e] choice and experimentation into New York City's school system" but warned it might "meet stiff resistance from the local educational bureaucracy."

When I learned that the KIPP schools were planning to open a charter school in Harlem, I proposed KIPP be placed in one of six underutilized buildings I'd identified. DOE's bureaucracy told Klein that I was misinformed and there was no excess room, but Klein, however, insisted on looking at the data himself and confirmed that there was. He placed KIPP at one of the schools I'd identified and began aggressively implementing the policy I'd recommended, which came to be known as "co-location."

In addition to running the Education Committee, I sought to get funds for nonprofits in my district, including several millions dollars for the Metropolitan Museum of Art, which I figured made up for coins I'd stolen from their fountain as a kid. I also helped Gifford in his efforts to manage the council. He often consulted me on ethical issues since he knew I had high standards in this area. For example, the chairman of the council's budget committee wanted to join a law firm that often lobbied his committee. He claimed it wouldn't pose a conflict because he'd recuse himself from any issue on which his firm lobbied. This was laughable. Even if the chairman didn't vote on the issue, the other committee members would want to please their chairman by making his firm's client happy. This, of course, was the whole reason this firm wanted to hire him. Gifford forbade the council member from accepting the offer. Through actions like this, Gifford raised the ethical standards of the council.

Several of our colleagues, however, were a real embarrassment to the city. One was Charles Barron, who held a reception for Robert Mugabe, a brutal dictator whose security forces regularly engaged in human rights atrocities. To Barron, however, Mugabe was an "African Hero" and he dismissed the human rights violations because "You didn't care about black Africans when whites were killing them." Many of my colleagues including Bill de Blasio went to this reception. Barron later said he sometimes felt that he wanted "to go up to the closest white person and say, 'You can't understand this, it's a black thing,' and then slap him just for my mental health."

Before all this, I'd taken Barron seriously. He'd given a speech in which he complained that all of the portraits in City Hall were of white men. This seemed like a fair point so I called Professor Kenneth Jackson, who was president of the New-York Historical Society, and asked him if he had any portraits of historically important African Americans that he could lend to the council. Sure, he said, so I scurried back to Barron to tell him the good news. Barron, however, was completely uninterested and I eventually figured out that he didn't want to address the problem, just complain about it.

On the personal front, I had become pregnant, much to my surprise given my prior difficulties, and on January 23, 2003, gave birth to a healthy redheaded boy whom we named Dillon. To avoid letting up on my work, I took him around with me as much as possible, and at my next hearings, as the *Times* reported, "Ms. Moskowitz held her 5-week-old son . . . in her lap, presiding with a burp cloth on her shoulder." Even more to my surprise, I became pregnant again just nine months later. It seemed like I was getting more fertile with age, not less. Go figure. When I went to the hospital to give birth, Culver said he hoped it wouldn't hurt when they cut open my belly, not knowing how babies actually come out, so I explained it to him. He responded, "You've got to be kidding," which is pretty much how most women feel about childbirth. I gave birth to a girl whom Eric and I named

Hannah. I was overjoyed since I'd long wanted to have three children and to have a daughter. Since Eric and I lived on the Upper East Side, we were among those fortunate New Yorkers who had good public school options. We sent Culver to PS 290, a progressive school with wonderful project-based learning that I later incorporated into the design of Success Academy.

While I'd generally been supportive of Bloomberg's efforts to improve the schools, I increasingly found myself in conflict with his administration. On January 3, 2003, the *Times* published a profile on me titled "City Council's Unapologetically Demanding Voice":

> It's no wonder, really, that Education Department officials refused to show up at City Councilwoman Eva S. Moskowitz's hearings.
>
> It cannot be a picnic to answer to the redhead in the red suit with the voracious appetite for data. Not easy to be told your department deserves an "F" for its school transfer policy or to be grilled about the status of capital projects . . .

While the administration chafed at my oversight, I felt this was shortsighted. It was my job as chair of the Education Committee to expose problems and build pressure for change. Besides, if I was perceived as being their lapdog, then my support for changes like mayoral control and imposing a standardized curriculum would have done the administration little good. Moreover, I had access to information that Klein didn't because DOE bureaucracy preferred to paint a rosy picture for him. My information came straight from unhappy constituents who told me what was really going on in their children's classrooms. For example, one constituent told me that the rug in her daughter's classroom was so dirty that students were getting rashes. It turned out that when DOE had decided elementary school classrooms should be outfitted with rugs, nobody had realized the custodians' contract didn't require vacuuming.

I also passed a bill requiring the mayor to report quarterly on the status of capital projects. When Bloomberg vetoed it, the *New York*

Sun editorialized: "Why the mayor would veto this bill—which simply requires the schools chancellor to report quarterly to the council on ongoing construction and repair projects, and notify it in the case of major delays or cost overruns—just as the state has given the council the responsibility to approve the school system's $7 billion five-year capital budget, is a puzzle." The council overrode Bloomberg's veto.

Another dispute I had with DOE concerned the federal No Child Left Behind Act, which gave parents the right to transfer students out of failing schools. Out of 220,000 students in failing schools, only 1,500 had gotten transfers. Many parents, my staff learned, hadn't been notified of their rights and many others who'd sought transfers had only been offered spots at other failing schools. Indeed, the system seemed designed to make it hard for parents to exercise their right to transfer. For example, Kamyah Harper, a mother in Queens, found out just days before the school year began that the school to which her son had been assigned, MS 198, had just been designated a failing school. She immediately called District 27 to get a transfer but was told it was too late even though she'd never previously been notified that the school was failing.

I asked DOE to testify at a hearing regarding failing school transfers and they responded that their job was "to educate children, not Eva Moskowitz." I have to admit that's a nice zinger, but DOE was in fact subject to the council's oversight so I threatened to subpoena the department. DOE agreed to testify and announced an overhaul of the transfer process a day before my hearing.

In practice, however, little changed so I wrote a letter to US secretary of education Rod Paige demanding the federal government enforce the transfer provision. He ignored my letter, probably figuring that a member of the New York City Council couldn't cause much trouble. The *Wall Street Journal* editorialized:

> If the Administration is alienating the likes of Ms. Moskowitz—a strong proponent of charter schools and public school choice—something is wrong . . . Ms. Moskowitz [wrote that] New York is

making "a mockery of the NCLB Act that [federal education officials] are charged with implementing and enforcing." Her letter is dated August 6, and she's still waiting for a reply.

Paige called me the next morning.

I had this faith that if I kept on pushing, the system would change, particularly with Klein and Bloomberg in charge. The more I looked into things, however, the more I became aware of just how dysfunctional and resistant to reform the school system was. Take a simple thing like trying to get in touch with your school. As a result of complaints from parents, I had my staff place calls to 110 schools between 4 p.m. and 5 p.m. At half of the schools, nobody answered; you couldn't even leave a message.

In addition, enormous sums were being wasted on constructing and renovating buildings. DOE was spending $438 per square foot to build new buildings, nearly twice as much as the $240 per square foot that Bronx Prep, the charter school Eric had helped start, had spent on its new building. Moreover, that greater expense didn't assure the work was done well. At PS 24 in Brooklyn, the building's entire facade had to be torn down and rebuilt just four years after the building opened. Renovations were also inefficient and expensive. At Curtis High School on Staten Island, for example, renovations had gone 455 percent over budget. One principal told me she was at her wits' end because, after years of delays, the city had finally gotten around to fixing a gym floor that had been damaged by a leaky roof but it had soon been ruined because DOE hadn't fixed the roof first. Another area of waste was scaffolding, which the law required for construction to protect pedestrians from falling bricks. While that made sense, DOE was taking so long to get around to fixing the facades that it was spending a fortune on scaffolding, which was rented on a monthly basis. About five hundred schools had such scaffolding. At PS 186 in Brooklyn, they'd spent $500,000 on scaffolding.

One of my colleagues got DOE to purchase air conditioners for PS 261 in Brooklyn, but they'd been sitting in storage for two years before the School Construction Authority got around to installing them. When I obtained funding for a music room at a school, I went to hear music played in the room and it sounded wonderful, which was great, but then I went to the classroom next door and the music sounded wonderful there too, which wasn't so great. The contractors had omitted insulation from the rooms figuring nobody would check. I made them redo it, but this type of shoddy work was no doubt overlooked 99 percent of the time.

The dysfunction of the school system also took a terrible toll on teachers. Some people believed that nobody wanted to become a teacher because it was a low-status profession, but I came across a lot of talented and energetic teachers on my school visits. The problem was that many were driven out of the profession or became demoralized. Right from the start, the system seemed to suggest that their time wasn't valued. For example, new teachers had to get fingerprinted. After getting complaints about the time it required, I sent a staff member to investigate. She had to wait four hours to get fingerprinted and anybody who arrived after 3 p.m. was told to return the following day. Teachers had to waste half a day on something that should have taken minutes. Moreover, this was just one step in a lengthy hiring process that involved visits to innumerable offices, and once they were hired, it often took DOE months to put them on the payroll.

Another problem was that the seniority system resulted in new teachers being assigned to the most difficult schools whose discipline problems they were ill equipped to handle. At JHS 226 in Ozone Park, teachers threatened to walk off the job when three of them were assaulted by students in a two-day period. At Norman Thomas High School, a student had punched a pregnant teacher in the stomach and said, "I'm going to kick the baby out of you. I'm going to make you have that baby." Instead of immediately suspending the student, DOE had allowed him to return to school until his suspension hearing took place. While I appreciate the desire to be mindful of student rights, you're going to lose your best teachers if you don't keep them safe.

I held hearings on a plethora of subjects including school lunches, facilities' maintenance, small schools, math instruction, education for incarcerated juveniles, bilingual education, gifted and talented programs, and school safety. But there was one topic that I'd shied away from: the union contracts that governed the schools' teachers, administrators, and custodians. I knew that taking on this topic might end my political career, and while I was willing to make that sacrifice, I wanted to be sure that if I did, I at least did this important topic justice. To do so, I needed to learn how to control committee hearings and handle difficult witnesses, to identify sources of information and expand my knowledge of the school system, and to develop relationships with reporters and earn a reputation for fairness. After two years as chair, I was finally ready.

30

THE TRIUMVIRATE

2012–2013

In just six years, Success had grown from one school with 275 students and a $4 million budget to fourteen schools with 4,000 students and a $72 million budget, and I was increasingly worried about my ability to manage it. Every year, it was a desperate race to get ready for the first day of school. Here's the report I gave to my board about what we'd done to open in 2012:

> [Renovated] Fourteen buildings—seven brand new—in thirty-six days
>
> [Hired] 310 new school employees
>
> Hosted a seventeen-day leader summit with seventy-plus school leaders
>
> Ran a fifteen-day [training] with 520 teachers
>
> Managed eleven major vendor groups, coordinating orders, delivery, and assembly of thousands of items (from principals' desks to pencils) to fourteen locations, up to the right floor and the right room, while navigating all renovations and summer school
>
> Defended five lawsuits/challenges to commissioner, two court appeals
>
> Moved our data center; collected, imaged, and assigned more than 1,200 laptops, 500 iPads; equipped every room with a SMART Board

Mac mini, and readied fourteen schools with wireless, printers, phones, and copiers

Welcomed 1,402 new families

While we'd managed to open, it had felt like we were on the precipice of disaster. It wasn't until days before school began that we'd renovated the last classroom and until weeks afterward that we'd hired our last teacher. My board felt we were lurching from one crisis to another: lawsuits, bad legislation, critical press stories, principals quitting at the last minute, and all the operational challenges created by our rapid expansion. Then a survey we'd commissioned found that while our employees had strong feelings of "personal accomplishment," many felt they lacked "work-life balance." My board feared I was running people into the ground. They also suspected that I was micromanaging my subordinates rather than delegating more responsibility to them as Success grew.

All of these concerns fed into what I would call the parable of the start-up, which went as follows. Start-up founders think out of the box, ignore naysayers, drive people to excellence, and micromanage. Long-term managers, however, must nurture talent, minimize risk, and delegate responsibility. The very qualities that make a founder successful often prevent her from becoming a long-term manager, but the founder resists bringing on such a manager because she loves her baby and doesn't trust anyone else to care for it—which, of course, is yet another symptom of those very failings that require her replacement. Some of the board members began to feel that maybe this parable applied to me, that I lacked the managerial chops to run an organization as large as Success. They proposed bringing on a CEO to take on the principal management function and elevating me to a founder/visionary/strategic leader role.

I was very conflicted. I did have concerns about my managerial abilities. I'd never worked in a large well-managed organization, been mentored by a strong executive, or studied business management. I was therefore learning on the job. For example, when we made a job offer to a new teacher, we'd often tried to match her current salary to get her to join but as we got larger, teachers began noticing disparities in their salaries we couldn't really justify. I learned that it's perilous to make ad hoc compensation decisions in a large organization. I had little experience dealing with these kinds of issues. I also worried I was losing control of the organization. Sometimes I'd find out a school was doing something crazy, I'd ask why, and the principal would say that it was a policy dictated by the Network even though I'd never even heard of the policy much less approved it.

So perhaps the board was right that I lacked the skills to continue managing Success. Moreover, this strategic-thinker role sounded attractive since I'd be spared the headaches of daily management. Another part of me felt, however, that this idea that I could just hand over everything to some brilliant manager was too good to be true. One reason we'd achieved so much at Success was that I'd rejected the orthodoxies to which others clung. I had refused to believe disadvantaged kids admitted through a lottery system couldn't be as academically successful as students in affluent suburbs. I feared a new CEO would be "realistic" about what our kids could achieve and that we'd end up with mediocrity.

I also felt that my board might be overestimating how hard I was making people work. My board members came from professions where eighty-hour workweeks were common so they probably assumed that if our employees were complaining, they must be working a hundred hours a week. In truth, nobody at Success worked as hard as big-firm lawyers or investment bankers. Sure, it wasn't a nine-to-five job, but we were seeking to revolutionize urban education and revolutions don't lend themselves to forty-hour workweeks.

But perhaps that was no longer a realistic way to look at things; perhaps I needed to accept that there was a limited supply of people who had the level of commitment I wanted and that my fear that letting someone else run Success would lead to mediocrity was just a symptom of founder's syndrome.

I didn't know what to do. I didn't want to make an error out of possessiveness or hubris but neither did I want to abdicate my responsibility out of timidity. I decided to see if I could learn more about managing a large organization, so I sought the advice of several people from the business world including one of my funders, John Fisher. He recommended I meet with Mickey Drexler, who had turned around Ann Taylor, played a big role at Gap, and now ran J.Crew. After exchanging a few pleasantries, Drexler pressed an intercom button that broadcast throughout J.Crew's entire headquarters and asked if anyone knew anything about me. Calls began pouring in and Drexler put each one on speaker, keeping my presence secret until the end when he took a mischievous delight in telling them I'd been listening to every word they'd said. (Their comments very much depended upon whether they knew somebody at Success or had just read about me in the papers. If the former, their comments were generally favorable; if the latter, not so much.)

I told Drexler I needed his advice on scaling Success without sacrificing quality. Drexler proceeded to grill me about my own thoughts on this. After doing so at length, Drexler finally appeared ready to provide me with the advice for which I'd come. To my surprise, however, he proceeded to tell me that I already knew how to scale. I told him I feared I was a micromanager who had unreasonable standards. Drexler scoffed at this, saying I wasn't half as bad as he was. He got involved in the smallest details of clothing design, personally interviewed every single person who was hired for the corporate headquarters, and visited J.Crew stores incessantly.

It was quite bewildering. I felt like Dorothy in *The Wizard of Oz* when the Good Witch of the North tells her that she'd had the power to get back to Kansas all along but just hadn't known it. But over time I thought about it more and began to feel greater confidence. While Drexler was extremely impressive, he hadn't appeared to have supernatural powers. Rather, he seemed to have many of the same qualities I had: high standards, attention to detail, and an anti-red-tape attitude as reflected in his unconventional use of his intercom.

I also sought the advice of one of my board members, Chuck Strauch, who had managed several businesses he'd started from scratch. I found Chuck to be wonderfully practical, wise, and kind. We were an odd couple—he's a conservative Republican from South Carolina and I'm a liberal New York Jew—but for some reason we hit it off instantly. Chuck felt that I was capable of running Success but that I needed a different organizational structure in which fewer people reported to me. He suggested I establish a triumvirate of three vice presidents who would report to me: Keri, who had impressed everybody with her abilities; Jody Friedman, our head of external affairs, who was exceptionally well regarded and more experienced than most of Success's employees as she'd worked for decades at Princeton; and a new person we'd hire who had enough business experience to help us adopt private sector best practices and deal with scaling issues.

I liked this plan and I felt increasingly confident that I could develop the skills I needed to continue managing Success if I put my mind to it. I realized, however, that I also needed to do a better job warning them about potential risks and sharing my strategic thinking. My failure to do that was giving them the impression I was running Success by the seat of my pants. I also needed to take better account of how they expected a CEO to act. Private sector CEOs typically exude self-confidence and make great efforts to convince their board they are doing a great job. My management style, however, was to be very critical both of myself and of others because I was setting a high bar. For example, if the board asked me how recruiting was going, I'd say that we were doing a horrible job of attracting candidates and I didn't know if we'd be able to hire enough. I realized I had to give them a more balanced view and do a better job explaining how we were going to handle the challenges we were facing.

I set out to right the ship in a meeting I had with my board in the summer of 2013. I presented to them the triumvirate plan Chuck and I had formulated and expressed my belief that I was fully up to the challenge of continuing to lead Success as its CEO. Fortunately, things had changed a little since the board had begun raising its concerns nearly a year earlier. By this time, Jody, who had a deft touch with powerful people from decades of working with strong personalities at the highest levels of philanthropy, had been with Success for eighteen months, and Keri, whom the board also greatly respected, had now been with Success for four years. Their presence and strong support for me helped combat the board's concern that I was a micromanager who drove away capable people. I also reassured the board that I understood the need to further strengthen our management team. While I didn't entirely alleviate their concerns, the meeting went reasonably well and they approved the triumvirate plan.

Given that things were now a bit more in control, I decided to do something in the summer of 2013 that I hadn't done in more than ten years: take a real vacation. I'd taken days here and there, but nothing like the long biking trips that Eric and I had once enjoyed. I'd been looking forward to the day when our kids would be old enough to join us on such trips and I felt that day had finally come, at least if Hannah rode on a tandem bike with Eric. We planned an ambitious monthlong trip in the south of France. While we'd have to make some concessions to being older and having children—no more camping out in fields—we still intended to do the trip on our own rather than as part of a group and to figure out our route on the fly rather than plan it in advance.

Eric had great fun showing the kids how to pack light. We took only one change of clothing that we planned to wash by hand daily and Eric was very strict: he inspected our bags before departure and upon finding any extra clothes or other contraband, issued a tongue-in-cheek reprimand. As a result, we each brought only a knapsack that was about the size of a basketball. When airport personnel commented that we weren't taking much baggage given that we were taking an international trip, Eric delighted in saying that this was because we were only traveling for a month.

We started our trip in Avignon and biked around five hundred miles in a big circle. We visited medieval castles, which our children particularly appreciated, and numerous Roman monuments including the arena at Nîmes; the ruins at Vaison la Romaine; the amphitheater at Orange; and the Pont du Gard, a huge bridge that was part of an aqueduct that once carried water twenty-six miles on a continuous downward slope from Uzès to Nîmes. It meant so much to Eric and me that we were able to share with our children this activity that had played such an important role in bringing us together in the first place.

31

180 MICHAEL BROWNS

2013

In April 2013, Randi Weingarten wrote to Daniel Loeb that she and "a small group of pension fund trustees," including "two funds that are current clients of yours," wished to meet with him. Weingarten had come to recognize that unions could use their influence over public pension funds to strong-arm investment managers like Daniel to stop supporting education reform efforts. Weingarten justified this because the money being managed ultimately went to benefit union members, but unions didn't really have any skin in the game: if pension investments performed poorly because managers were selected on grounds other than their investing abilities, taxpayers had to pick up the bill, not unions. While Weingarten's tactics undoubtedly cowed some investment managers, Daniel didn't take well to being bullied. "It's important to make a determination," he said, "what's more important to you, your money or your principles? For me, it's my principles." Thus, rather than pulling back, Daniel doubled down and agreed to become chairman of Success's board in October 2013.

Within the Network, we had an increasingly robust management team including Chief Academic Officer Michele Caracappa; Managing Director of Schools Jackie Albers; Managing Director of Science, Math, and Technology Stacey Gershkovich; and Managing Director of Elementary Schools Paola Zalkind. They are all extraordinary educators and leaders and were in large part responsible for the fact that our academic program had actually improved over time, not worsened as I'd feared might happen. Remarkably, three of the four had been among our very first crop of teachers.

However, there were areas where I felt Success was falling short of the ideal that I'd imagined when I'd founded it. One was athletic competition, which provides opportunities for parents to root for and take pride in their children and for students to develop confidence in themselves and camaraderie with their peers. This had certainly been true for Culver, who'd earned the nickname "Big Shot Moskowitz" for his basketball skills. While our students had played soccer since kindergarten, they weren't progressing. They'd crowd around the ball, kicking at it wildly but with little passing or teamwork. I feared the AstroTurf fields we'd built were too small. We considered bussing our kids to larger fields but that would be expensive and would take precious time away from academics. We presented our problem to a new coach we'd hired, Boris Bozic, who'd played professional soccer in Serbia. It turned out we weren't the first ones to grapple with this problem and Boris had a solution.

In the 1930s, a devout Catholic in Uruguay by the name of Juan Carlos Ceriani Gravier wanted the children in the YMCA program he ran to play soccer but, like us, lacked proper fields. He therefore modified the game so it could be played indoors. He decreased the size of the goals and the number of players, and used a heavier, smaller ball that went less far and bounced less high. Gravier's invention, which came to be known as "Futsal," spread throughout South America and revolutionized soccer. The fancy footwork and precise passing Futsal required would give rise to a new style of play that was more flamboyant and fast paced, exemplified by players like Messi, Ronaldo, Neymar, Zico, Sócrates, and Pelé. "Futsal makes you think and play fast," said Pelé. "It makes you a better player."

Not only would Futsal improve our students' soccer skills, Boris explained, but our small AstroTurf fields and indoor gyms were perfect for it. We could have a world-class soccer program without wasting countless hours and spending loads of money bussing our kids to distant soccer fields. Boris created a program for Success that began in kindergarten with exercises such as kicking balls around cones to develop foot skills and agility. In third grade, students who showed the most commitment and talent would be selected to participate in a competitive soccer program, although all students who wanted to continue to learn soccer would have an opportunity to do so. Thousands of kids now participate in our soccer program.

In addition to soccer, I wanted to provide more opportunities for children to pursue their interests so we introduced an electives program for our middleschoolers. Each day began with an hour-long period in which students could choose classes in art, soccer, volleyball, musical theater, chess, entrepreneurship, newspaper/media studies, creative writing, or debate.

We also tried to do a better job of challenging our strongest students. When educators say that "every child" should be allowed to fulfill his potential, they often really mean "average and struggling students." Every child should mean every child, including gifted children. They have just as much right to reach their potential as do struggling children. It's easy to miss when strong students aren't being challenged. Tests help identify struggling students, but if a gifted child aces a test, educators usually just pat themselves on the back. They don't ask themselves how much of the material the student already knew or how much the student could have learned if he'd been challenged.

We started administering pretests to find out which students already knew what we were about to teach them. Once we identified gifted students, meeting their needs wasn't easy, but we started giving more advanced students different homework or additional problems to do in class. We also gave kids opportunities to do independent work such as teaching themselves from books like the *Challenge Math* series by Ed Zaccaro.

We also pushed our teachers to increase the rigor for every student. Many teachers like to make things easy for kids. They come up with nifty little mnemonic devices that help students solve problems without really understanding them. They give kids sample questions that are nearly identical to questions that appear on tests. They use the simplest words possible when speaking to children for fear that some child won't know a word they are using. They do all this with the best of intentions, figuring that the easier they make it for kids to learn, the more they will, but cutting up things into bite-size pieces often impedes children's ability to learn how to tackle difficult problems.

Once, a girl who was struggling over what move to make next told Sean O'Hanlon that chess was hard. "Yeah," he replied, "it's supposed to be hard." So is school. Intellectual struggle builds a child's intellect just as lifting weights builds an athlete's muscles, and children can do more than we think. At the height of their language acquisition, children learn seven words a day and that happens mainly just by hearing them used. If you use only the words a child already knows, you're depriving him of the opportunity to learn new ones. We therefore asked our teachers to use more advanced vocabulary and to pitch their instruction at a higher level.

As the school year came to an end, I needed to figure out who I'd make principal of a new school we were opening in Harlem. Khari had long been gunning for this promotion and I felt he was now ready and was glad to give him this opportunity. For a midlife career changer like Khari, becoming a principal in the district schools would have been extremely difficult given the cost and time required in getting a principal's license. Being able to shortcut these unnecessary formal requirements allowed him to succeed professionally and allowed us to access this nontraditional source of talent.

A year after I made him principal, an incident occurred that confirmed Khari's commitment to our mission. He called to tell me he might miss the first day of school to join protests in Ferguson, Missouri, over the police shooting of Michael Brown. I thought it was more important that he be at his school, but the decision was his to make, so I just told him that if he got arrested, he should give me a call so I could make sure he had a good lawyer.

On Sunday night, he sent an email to his faculty explaining his decision:

> Tomorrow morning, we will open the doors to another school year and greet over 180 children. Many of those children will look very much like Michael Brown did on his first day of school. My job is to be there for them. It is our responsibility to build an environment where children get the education they need to change the world, be power brokers, build institutions, and improve upon the design of a country where their lives—and the lives of their progeny—are protected, valued, and respected. Tomorrow, when our doors open, we will be standing against educational injustice. Tomorrow, there will be 180 Michael Browns walking through the door. Tomorrow begins the day we continue to get the Michael Browns of Harlem to college and beyond and KNOW that they will be prepared to engage the world on their terms.

Following this was a page or so of quite specific directives by Khari to his teachers to make sure that school started well.

Khari saw his work as a means of advancing social equality but understood the hard work that it truly requires. It's not enough just to teach children about the civil rights movement and our country's oppression of minorities. Knowing about past injustices won't help children succeed in the future unless they also have the requisite academic skills, and teaching them requires hard work. It requires continually striving to improve, connecting with children emotionally, supporting one's colleagues, always being punctual, and giving 100 percent day in and day out.

Fortunately, the opposition to Success had temporarily subsided but I had no illusions that we were out of the woods. The vitriol that some felt for us would bubble to the surface every now and then. An article in the *Village Voice* quoted one critic in Brooklyn as follows:

> What the f_ _k? Who the hell are you? How do you get to decide we need a new school? . . . I get that Eva Moskowitz is sociopathic enough to put her own kids in this school to prove a point . . . [b]ut [n]o person of means would.

Incredibly, the article's author was actually sympathetic to this woman and claimed that I was "trampl[ing] parents' rights" by offering them the option of sending their children to our schools and that I was a "carpetbagger" because—I kid you not—I wasn't from Brooklyn.

This continuing undercurrent of hostility to charters worried me because there was a looming danger on the horizon: we were about to lose our most important supporter.

32

TIME FOR EVA MOSKOWITZ TO STOP BEING TOLERATED, ENABLED, SUPPORTED

2013

The year 2013 was the final year of the reign of Mayor Michael Bloomberg, the business titan who, like Plato's philosopher king, had descended from his world of fathomless wealth and privilege to become "Mayor Mike" and govern over the shadowy affairs of municipal government. Over the years, our paths had intersected many times. Bloomberg had been one of my earliest campaign contributors and had served as mayor for most of my tenure on the council and all of the time I'd run Success. While we'd sometimes been at odds, we'd worked together more often than not, and Success's growth was in large measure due to his unflagging support for charter schools and co-location.

Given how much I'd benefited from Bloomberg's policies, some assumed we were best buddies. In fact, we never hit it off personally. Bloomberg is from Boston and perhaps I was a bit too much like the New York stereotype for his taste: impatient, brash, aggressive. For my part, I sometimes found him diffident and lacking in humility.

But that's personalities. Substantively, I have nothing but respect for Bloomberg. Many people are in politics for fame, adulation, or power. Bloomberg didn't care about any of that. He didn't savor being called "Mr. Mayor" or being applauded or handing out awards. He didn't seek adulation but instead discouraged it. Toward the end of his time in office, he explained his view:

> If I finish my term in office . . . and have high approval ratings, then I wasted my last years in office. That high approval rating means you don't upset anybody. . . . Well, you're skiing the baby slope, for goodness' sakes. Go to a steeper slope. You always want to . . . tackle the issues that are unpopular, that nobody else will go after.

Why would someone who didn't care about being popular spend a half billion dollars of his own money so he could be mayor? There was only one reason as far as I could see: he loved the work. He believed there were important public policy issues facing the city and that he was the best qualified of its eight million citizens to address them; which of course was arrogant, but nonetheless correct. He left New York City a far better place than he'd found it not by advancing fundamental ideological change, as Giuliani had, but through better management. He focused on easing traffic, fixing the economy, improving the schools, decreasing smoking, beautifying the parks, and fighting crime. His administration planted more than 800,000 trees, turned a $6 billion deficit into a $3 billion surplus, and cut the murder rate in half. Indeed, on November 26, 2012, something happened in New York City for the first time in recorded history: the city went a full day without a single murder, shooting, or stabbing.

While I was disappointed Bloomberg hadn't taken a harder line on the UFT contract, he probably correctly judged that it just wasn't politically feasible. I was more of a purist, but you know how that turned out. If Bloomberg had taken a similar route and lost his election, he wouldn't have been around to continue the co-location policy that was so vital to the proliferation of charter schools. For me to be me, I needed him to be him.

I met with Bloomberg in the waning days of his administration and he seemed dispirited. The life expectancy for a man his age, he observed, wasn't long. I was surprised that he was confiding in me and even more so by what he said next: that he doubted public education could be fixed in what was left of his lifetime. That one person could even think in these terms—of fixing education, rather than simply improving it or leaving one's mark on it—reflected the scope of his ambitions. When most of us think about death, we worry about the winking out of our consciousness forever, about what, if anything, lies for us beyond this world; Bloomberg seemed more worried about what lay ahead for the world beyond Bloomberg, how we'd manage the problems of public education, health care, and gun control once he exited stage right.

They say of a great athlete that he is "playing a different game." I've tried my best to contribute toward improving our city and I'd like to think I'll leave my mark—but Bloomberg played a different game. He'll go down in history not only as one of New York City's greatest mayors, but one of our country's greatest citizens.

I gave some thought to running for mayor myself, as I'd long been interested in the job and felt I had a shot at winning, but I didn't feel I'd gotten Success to a point where I could leave it. Of the candidates seeking to succeed Bloomberg, my favorite was Chris Quinn, who'd worked with me to help Gifford Miller become speaker and now held that position herself. She was pragmatic, intelligent, ethical, and supportive of charter schools. She had four opponents, two of whom particularly worried me. One was my former colleague Comptroller John Liu. While I liked him personally, he wasn't supportive of charter schools and his campaign treasurer had been convicted for accepting improper campaign donations. But the real nightmare scenario for Success would be the election of my former colleague Bill de Blasio, who claimed he wasn't anti-charter but somehow ended up opposing every policy that would help us and supporting every one that would hurt us. As public advocate, he'd put out a joint report critical of co-location with the Alliance for Quality Education (AQE), a union-funded group. For a government agency to issue a joint report with a private advocacy group is just bizarre. Imagine if the Justice Department put out a joint report with the American Civil Liberties Union or the National Rifle Association. It just isn't done, and the fact that de Blasio would so nakedly turn his office into a mouthpiece for union propaganda was deeply troubling.

Moreover, it became increasingly apparent that de Blasio was not only anti-charter but anti–Eva Moskowitz. At a mayoral debate, he announced that, "Another thing that has to change starting in January is that Eva Moskowitz cannot continue to have the run of the place." Then, on April 24, he held a press conference in which he claimed that Success had received favorable treatment because some of the fluorescent light fixtures at our Cobble Hill school had been replaced while those at the co-located school hadn't.

"Time and time again," he said, "we've seen a *Tale of Two Cities*, with resources lavished on Success Academy while traditional public schools in the same building lacked the most basic necessities."[24]

This claim was utterly false. First, the city hadn't "lavished" funding on us; we'd paid to replace our own light fixtures. Second, not only did the co-located district school have "basic necessities," the city had actually spent $2 million on our co-located schools to give them "new wiring, locker rooms, [and] a dance and fitness center" for this school, "as well as creating and upgrading classrooms for students with disabilities," because of the state law requiring matching expenditures on district school facilities.[25] And finally, the district schools' light fixtures were scheduled to be replaced that summer anyway.

This was a hot-button issue because the light fixtures contained PCBs, a health hazard. De Blasio was implicitly claiming that DOE was putting the health of our children ahead of those of district school students, an incendiary charge. In fact, the risks had been overblown and had nothing to do with why we'd replaced some of our fixtures. We'd done so only in hallways that were being renovated and purely for aesthetic reasons; if safety had been our concern, we'd have replaced the classroom fixtures as well. Neither was it true that DOE was prioritizing replacing our fixtures. Of our fourteen schools, eleven had been treated exactly the same as their co-located district schools: either none of the fixtures in both schools had been replaced (six buildings), all had (three buildings), or some had (two buildings). And our three other schools had been treated worse: none of their fixtures had been replaced, while some of those in the district schools had.

But de Blasio insisted I was getting special treatment, claiming: "It's time for Eva Moskowitz's privilege and power to end."[26] One would think "privilege and power" would be something snazzy like the use of the city's police helicopters or a wing at Gracie Mansion, not a bunch of fluorescent light fixtures.

Eager to get in on the Eva bashing, my former colleague John Liu seized on an opportunity when a light fixture at PS 123 malfunctioned. An expert hired by DOE later concluded that the fixture had "overheat[ed], causing a burning or smoking odor" and that no toxins whatsoever had been emitted,[27] but Liu and Robert Jackson demanded "an immediate and thorough investigation into legal and environmental violations associated with toxic PCB lights." "Two schools," they claimed, "have had environmentally hazardous incidents with PCB lighting" and "Both schools are co-located with schools run by Eva Moskowitz's Success Academy." This made it sound like I was somehow responsible for environmental violations. In fact, the incident at PS 123 had nothing to do with me or with Success nor was it even an "environmental hazard," and there had been no incident at all at our Cobble Hill school. Nonetheless the news blog *DNAinfo* reported that "a burst light bulb released a toxic cloud of chemicals Tuesday and... caused the school to be evacuated." A "toxic *cloud*"! Evacuation! The absurdity just grew. Suddenly, it was Chernobyl in Harlem!

In May, the UFT held a mayoral forum that one newspaper described as a contest among the candidates to "prove that they despise former Council member Eva Moskowitz even more than the UFT does."[28] When asked what *Times* columnist Michael Powell called the "would you toss Eva Moskowitz into the dragon's mouth question,"[29] mayoral candidate Anthony Weiner (yes, he of selfie infamy) was so exasperated at the silliness of this exercise that he commented sarcastically, "I have no bloody idea. . . . Uh, sure. . . . It seems to be the answer of the day."[30] The question, however, was right in de Blasio's strike zone. He proudly declared that it was "time for Eva Moskowitz to stop having the run of the place. . . . She has to stop being tolerated, enabled, supported. . . . It wouldn't happen if she didn't have a lot of money and power and political privilege behind her, and if DOE didn't say, 'Yes, ma'am,' every single time. That's going to end when I'm mayor."[31]

Call me silly, but isn't saying someone shouldn't be tolerated a little, well, intolerant?

Then, de Blasio yet again invoked my name at a forum in June, declaring that, "There's no way in hell Eva Moskowitz should get free rent." You may wonder whether I was shacked up in some Fifth Avenue palazzo at taxpayer expense. Not quite. De Blasio was talking about the fact that Success, like all other district and charter schools, doesn't pay rent on the public school facilities it uses.

De Blasio had now criticized me at least four times in public forums, usually with intemperate language like "hell," "tolerated," and "yes, ma'am" thrown in. There are eight million other private citizens in New York City, including slumlords, drug dealers, insider traders, murderers, rapists, tax cheats, Ponzi schemers, and embezzlers, but I was the only one whom de Blasio had repeatedly singled out by name. I was an educator of primarily poor and minority kids, but somehow I'd become public enemy number one.

But what troubled me more than de Blasio's personal attacks was his proposed policies. He wanted to charge charter schools rent because, "There are charters that are much, much better endowed in terms of resources than the public sector ever hoped to be." In fact, that generally wasn't true, and it certainly wasn't true of Success: we deliberately spent no more per pupil than the district schools. The money we raised went to pay the start-up costs for our new schools: for furniture, books, technology, renovations, and to cover the operating deficits these schools ran in their first few years. De Blasio's claim was particularly ironic because his own son attended Brooklyn Tech, a district school with an $8.7 million endowment and a student body selected by means of a rigorous test. But de Blasio didn't think his son's selective school with a generous endowment should pay rent, only a charter school serving primarily disadvantaged students selected by lottery.

After the mayoral debate in May, the *Times* editorialized that "Shoehorning [charter schools] into existing school buildings over local objections can alienate parents and reinforce among students a harmful sense of being separate and unequal." I couldn't believe it. The phrase "separate and unequal" was the battle cry of the most extreme opponents of charter schools. Great, I thought; most of the candidates were already vowing to restrict co-location to suck up to the UFT, and now the *Times* was egging them on as well. Fortunately, although bizarrely, the *Times* reversed itself just two months later, editorializing against "a proposal by the teachers' union that would give a local community panel veto power" over co-locations.

As summer began, Weiner was leading the pack with 25 percent, followed closely by Chris with 20 percent. De Blasio had only 13 percent and the UFT's endorsement had gone to another candidate, Bill Thompson, despite de Blasio's promises to do everything to me short of driving a wooden stake through my heart. Even the UFT, it seemed, recognized that de Blasio was too far to the left to get elected in this day and age. Then Weiner's candidacy was derailed by his sexually explicit text messages and the polls soon showed that Chris was in the lead with 27 percent.

Yet Chris didn't seem to be really clicking with the voters. People had trouble relating to her, and Democrats who had Bloomberg fatigue felt she was too close to him. De Blasio's numbers started inching up, and on August 4, the *Times* published an article on de Blasio that began as follows:

> In a mayor's race crammed with celebrity razzle-dazzle, historic candidacies and tabloid turns, a gangly liberal from Brooklyn is quietly surging into the top tier of the field by talking about decidedly unglamorous topics: neglected hospitals, a swelling poverty rate and a broken prekindergarten system.

The article might as well have been an endorsement. It touched all the right points. "Gangly" and "quietly surging" were brushstrokes evoking the prototypical strong but soft-spoken leader (think Lincoln), while "unglamorous topics" rather than "razzle-dazzle" suggested he was substantive and concerned with real issues. In fact, the "unglamorous topics" the *Times* listed were anything but. Repairing infrastructure, balancing the budget, and reining in pensions are the types of things that must be done but that don't earn you points with voters. De Blasio's platform was precisely the opposite: a grab bag of goodies for people whose support he wanted. His position on hospitals appealed to New York's most influential union, 1199, which endorsed him; his position on pre-K appealed to voters with young children; and his promise to redistribute the wealth appealed to all of the voters who figured they'd be on the receiving end of the redistribution. Finally, the article even featured a picture of de Blasio in church (church!!) looking adoringly at his African American wife. If de Blasio had staged the picture himself, it couldn't have been better.

Meanwhile, Chris seemed paralyzed by the fear that she'd lose her lead. She was an intelligent, funny, and warm person but none of this was coming across to voters. It saddened me both as a friend who wanted her to win and as a citizen who wanted her to be my mayor.

With de Blasio's numbers growing, the press finally started to take a more critical look at him but he'd peaked at just the right moment. It was too late for the press to dig up dirt, for his competitors to do much damage by refocusing their firepower on him, or for the voters to have second thoughts. Now in the lead, de Blasio didn't back off of his anti-charter hostility an iota. To the question "Should charter schools be supported and increased?" de Blasio had responded, "No. We don't need new charters."[32] Then, on September 3, de Blasio put out a press release that called for "an immediate halt to co-location . . . plans for the remainder of Bloomberg's term," claiming that the co-locations being considered would result in "overcrowded" schools and "larger class sizes." De Blasio also objected to Bloomberg approving co-locations that would take effect after he left office although this was no different from bidding out work on a tunnel that would begin in the next mayor's term. You can't put all government business on hold during the last year of a mayor's term or government would grind to a halt every four years.

On September 10, de Blasio won the Democratic primary with just over 40 percent of the vote and it was clear that the Republican candidate had no chance. My nightmare scenario had come to pass. New York City was about to elect a mayor who was anti-charter, anti-Success, and anti-me. More important, the proposals he'd made—limiting future co-locations, revoking past ones, and charging rent—would be disastrous for charters and the students they served.

For many years, the charter movement had been lucky. While much of the political establishment was hostile to charters, when it came to the top positions of governor and mayor, the pro-charter candidates had somehow always ended up on top: Bloomberg, Giuliani, Pataki, Spitzer, Paterson, and Cuomo. I'd long known that we had to prepare ourselves for the day when our luck ran out and we had to stand on our own two feet. That day had now come.

In my career in politics and education, I'd seen a lot of action: I'd been written about hundreds of times, protested, sued, attacked in the media and in mailings, and fought pitched battles with political opponents, regulators, and unions. But it would all pale in significance compared to the maelstrom that engulfed me when de Blasio took office.

33

HOW MANY UNIONS DOES IT TAKE TO SCREW IN A LIGHT BULB?

2003

On July 23, 2003, I was on the floor of the city council chambers when several shots rang out. Gifford's bodyguard, who was standing next to me, dropped to his knees, pointed his gun up toward the balcony, and fired. Terrified, I dove under a table. Thoughts raced through my mind. Was this another terrorist attack?

After what seemed like ages, the police told us everything was safe and I crawled out from underneath the table. I soon learned that James Davis, one of the candidates I'd helped elect, had been shot by the candidate I'd helped him beat, Othniel Askew. Sadly, Davis died before medics arrived, as did Askew, whom Gifford's bodyguard had shot. We later learned that Davis had brought Askew to city hall to honor him for his work. Ironically, Askew had managed to get a gun into city hall because Davis had escorted him into the building and they'd been waved past security.

When I recovered from this disturbing incident, I returned to my work on the hearings I was planning to hold on the union contracts for the school system's teachers, administrators, custodians, paraprofessionals, and guidance counselors. These contracts determined virtually every aspect of how the city's schools were run: how employees were hired, fired, compensated, promoted, and supervised; what each type of employee could and could not be required to do; how much money was spent to maintain each school building; how often parent-teacher conferences could be held; where a teacher could be assigned; how often a principal could go into her classroom; what subjects she was allowed to teach; and the number and length of breaks to which she was entitled. Yet, despite their importance, these contracts were little understood. In part, that was because they were abstruse. The teachers' contract alone ran 77,841 words, which is 4 times as long as Shakespeare's 154 sonnets, 5 times as long as Albert Einstein's paper on the general theory of relativity, 10 times as long as the United States Constitution (including all 27 amendments), 59 times as long as the Declaration of Independence, and 286 times as long as the Gettysburg Address. Placed end to end, its pages would be taller than the United States Capitol Building and about two-thirds the height of the Washington Monument.

But the primary reason nobody had scrutinized these contracts was that the unions didn't want them to, and I knew my decision to do so could have profound consequences for my career. While I'd annoyed the teachers' union, Weingarten had been sending out signals that she was willing to let bygones be bygones. She'd praised my hearings on mayoral control and observed that it was "interesting" to watch my views "evolve." In other words, just "evolve" a little bit more and all will be forgiven. Plainly that would be the smart career move, but I didn't see the logic of advancing a career in public service at the cost of serving the public.

To help the public understand these lengthy and obscure contracts and why they mattered, I'd first have to understand them myself. My academic background again came in handy, as did the assistance of a brilliant and hardworking attorney by the name of Mark Goldey whom I'd recently hired for the committee. After studying the language of these contracts for six months, we sought to understand their real-world impact by speaking with superintendents, principals, and teachers. Finally, we needed some way to summarize what we'd learned and, realizing that we essentially wanted to create a CliffsNotes, we created booklets in the same style and called them "Council Notes."

We decided to hold one day of hearings on each of the three main contracts—those for the custodians, teachers, and principals. The most important one, we believed, was the teachers' contract, and the most relevant witnesses to testify about its impact were principals, since its provisions dictated how they managed teachers. While many principals had readily confided in us that the teachers' contract interfered with their management of teachers, most were petrified about saying so publicly for fear of retribution from the teachers' union. Mark suggested we record interviews with them and play them back with sound distortion. Three principals agreed to testify in this manner and two more were brave enough to testify live.

I kept my plans secret as long as possible since I feared that when word got out, the unions would try to stop the hearings. Sure enough, as soon as I went public, elected officials began calling me to cancel them. Even a United States senator called me. Only one elected official expressed support: Congressman Charles Rangel of Harlem, who told me it was about time someone stood up to the teachers' union. While I appreciated his providing this encouragement privately, it wasn't exactly comforting that he was unwilling to do so publicly despite having held his seat for decades.

A couple of days before the hearings were set to begin, one of the principals who'd agreed to testify live canceled because he'd gotten intimidating phone calls, including ones from his own union. I begged him not to pull out, but it was to no avail. I began panicking. The most important witnesses in the entire hearings were principals and now I had only one left who was willing to testify live. I feared he'd pull out too and I'd be left with nothing. Although I have an independent streak, it was all beginning to feel like too much. My witnesses were being intimidated and the entire Democratic political establishment was closing ranks against me. I'd been willing to risk my political career for these hearings, but if they fell apart, and I failed to educate the public, it would all be for naught. My failure would just embolden the unions and reinforce their power. Maybe this time, I'd really overreached.

But it was too late to turn back now. When you have lemons, you make lemonade, so on November 10, I held a press conference to make public these efforts to stop the hearings. The following day, the *Times* printed an article titled "School Unions Want to Cancel Labor Hearings, Official Says." It read in part: "The chairwoman, Eva S. Moskowitz, also said that many potential witnesses were too scared to testify, describing an atmosphere of fear that she said brought to mind Frank Serpico, the whistle-blower on police corruption who testified for the Knapp Commission in the early 1970's." With an opener like that, at least people would be paying attention to the hearings. It might still be the end of my political career, but at least it wouldn't be for nothing.

The hearings began on November 12. Our first witness was a DOE official who testified about a curious provision of the custodians' contract called "retainage":

> MARTIN OESTREICHER: *[The custodians] get an allocation . . . and then they have to spend a significant amount of money to maintain the building. [I]f you subtract one from the other, that could be their retainage . . .*
>
> MOSKOWITZ: *So after the custodian has paid his or her employees, they are allowed to retain up to a certain level [of] monies; is that correct?*
>
> OESTREICHER: *Right.*

Thus, custodians had an incentive to spend less maintaining their buildings so they could keep more. Custodians could also increase their compensation, Oestreicher explained, by getting assigned to a second school that was temporarily without a custodian. In theory, that required more work, but in reality, according to the principals who testified, custodians would just split up their existing work hours. Through these various manipulations, custodians could make up to $184,000 annually. By contrast, a principal, who managed far more employees and was responsible for the education of hundreds of students, could make only $115,000.

Oestreicher also testified about how a school's custodial budget was set. Any home owner knows that the cost and difficulty of maintaining and cleaning a house is affected by many factors, such as how old it is and how many people live in it, so I asked Oestreicher how these factors influenced a school's custodial budget:

MOSKOWITZ: *Does [it] take into account the age of the building?*
OESTREICHER: *No.*
MOSKOWITZ: *And does a 100-year-old building require more maintenance?*
OESTREICHER: *Yes.*
MOSKOWITZ: *Okay. And the condition of the building is also not a factor?*
OESTREICHER: *No.*
MOSKOWITZ: *If a building [is] at let's say 140 percent capacity, would [it] require more maintenance than a building at 100 percent capacity?*
OESTREICHER: *Yes.*
MOSKOWITZ: *And is the allocation any different?*
OESTREICHER: *No.*
MOSKOWITZ: *Why is it that we have a system that doesn't take into account what seem like fairly important factors?*
OESTREICHER: *The factors are determined by the contract. [I]t's not something that the Department can determine itself.*

While I understood why a labor contract would determine a custodian's salary, I didn't understand why it should determine a school's maintenance budget. It was as if an airline determined how much it spent on maintaining each airplane based not on how much maintenance it needed but on the terms of the pilot's contract.

The contract also had very little by way of minimum requirements for custodians. While an appendix said they should "wash and disinfect all toilet seats" and "fill all soap dispenser systems," it expressly stated that "these are suggestions only."* Moreover, the contract often limited what they were even allowed to do. For example, custodians couldn't change the ballasts that made fluorescent lights turn on:

> PRINCIPAL: *[The custodian] could change bulbs but [not] ballasts. . . . And the [DOE] won't send an outside electrician to do it unless you have something like 50 ballasts out. . . . [I]t becomes a nightmare because then you have kids sitting in the dark.*

The result was often finger-pointing:

> MS. DEANGELIS: *[The custodian would] tell me the ballast is out and we can't do anything about it. . . . But [some custodial assistant workers] told me that some of them . . . were light bulb problems, because in the evening they would change some of the light bulbs and get them working.*

In other words, the answer to the question "How many unions does it take to change a light bulb?" was "two," which coincidentally was also the answer to the question "How many unions does it take to paint a classroom?":

> OESTREICHER: *The ten-foot rule [says] the custodians can paint up to ten feet and above that, that work must be turned over to the painter trades.*
>
> MOSKOWITZ: *So he would be violating the contract if he were to paint eleven feet?*
>
> OESTREICHER: *Yes.*
>
> MOSKOWITZ: *How does this work? Does all the furniture get moved into the middle, the custodian does ten feet, and then you move it*

* http://www.nysun.com/editorials/a-mess-to-clean-up/77974/.

back and wait for the painters, or do they come at the exact same time?

OESTREICHER: *[I]t's not all as coordinated as we'd like it to be where the custodian will move every desk into the middle, go ten feet, and there's a painter standing right next to him who says I'll take it from here.*

MOSKOWITZ: *Have you ever had it work seamlessly, where the custodians come in and they paint and then the painter is right behind?*

OESTREICHER: *I have never seen that.*

I asked Oestreicher how this crazy system had come about, and he explained that the custodians' and painters' unions had gotten into a fight over who had the right to paint in schools so "a determination was made [in] an arbitration." In other words, the ten-feet rule had been created by an arbitrator in a dispute between two unions without anybody looking out for the best interests of children or taxpayers. It was as if King Solomon had proposed splitting the baby and the women had said, "Works for us!"

My next day of hearings was on the teachers' contract. This would be the real showdown since the UFT was by far the city's most powerful union. Randi Weingarten had been waffling about whether she'd testify. Doing so would give credence to the idea that the contracts were an appropriate subject of inquiry, which she disputed. Refusing to testify, however, would leave the union's viewpoint unrepresented. At the last minute, we were informed she'd testify, which I took as a good sign since it meant she'd concluded our hearings had gotten too much traction to ignore.

When she showed up the following morning, she arrived with a man who sat down beside her in the front row and who, when I later called upon Weingarten to testify, joined her at the witness table where he fixed the committee members with an icy stare. Without saying a single word, he had more influence on these hearings than any other person. He was Brian McLaughlin, the leader of New York City's Central Labor Council, an umbrella organization for more than 400 unions with more than 1.5 million members, and was also a member of the State Assembly. But what truly made him powerful was not merely the positions he held but his willingness to use them without regard for ethics or legality. Of course, he hid his misdeeds behind closed doors, but they would be brought to light two years later when he was indicted and pleaded guilty to stealing over $2 million from taxpayers, labor unions, and not-for-profits including a Little League baseball team. Among his schemes was demanding that a company that had gotten a contract to upgrade the city's traffic lights give him $450,000 to avoid labor trouble. He used his ill-gotten gains to pay for luxuries such as a country club membership, a plasma-screen television for a girlfriend, and an $80,000 Mercedes-Benz. At the time of my hearings, McLaughlin had reached the peak of his power and was even rumored to be considering a run for mayor.

McLaughlin later explained that he was at the hearings "to remind the city council members that the entire labor movement in the city is watching them." In other words, there would be consequences if they joined my efforts to examine the terms of the labor contracts. Weingarten would be forever grateful to McLaughlin for helping keep my colleagues in line. When his corruption came to light, she said she was "rooting for"[33] him because "even if you're in a fight that's unpopular"—an obvious reference to my hearings—"he'll stand with you."[34]

The first witness that day wasn't Weingarten, however, but the only principal who was brave enough to testify openly, Tony Lombardi, who described his frustrations with how teachers used the sabbaticals that were supposed to be for professional development:

> I have known teachers who have taken courses such as swimming, golf, history of the movies and contemporary cinema, human sexuality, and TV communications. . . . Courses that discuss Fellini's use of montage in his early films do not help my students meet academic standards. . . . You as citizens . . . were paying the teachers 70 percent of their salary to take these courses . . .

Moreover, teachers could take these sabbaticals whenever they wanted, which created scheduling nightmares. One spring, five of Lombardi's teachers had taken half-year sabbaticals simultaneously, so he'd been forced to cover their classes with day-to-day substitute teachers for the remainder of the year.

Lombardi also objected to a provision of the contract that prohibited him from collecting lesson plans from his teachers. One of his teachers wrote his lesson plans on napkins. Lombardi explained that

> planning is the heartbeat of teaching. . . . You can't just randomly teach whatever you feel like teaching each day. So what I teach today has an impact on what I teach tomorrow and the next day and the next day.

Even more problematic were the limitations on hiring and firing teachers. Principals could choose between two systems for hiring teachers. One required that 50 percent of vacancies be filled based on seniority. As Lombardi explained, "A teacher will just send me a letter saying 'I have twenty-five years' and . . . that person gets the position." The problem with this was not just that this teacher might be weak but also that she might be ill suited to the school since different schools have different needs. Some schools, like Stuyvesant, need teachers who can teach calculus or AP physics; others need teachers who are strong in remediation or behavior management. Schools may also be premised on different pedagogical beliefs. None of that, however, could be taken into account.

The second alternative was the "school-based option," a process in which candidates were selected by a committee, most of whose members were teachers. One principal described it as follows:

> The process itself is so cumbersome and so time consuming [and] the rules and regulations are so extensive and so potentially litigious that everything that one does as a Committee is based on this

> possibility of grievance. . . . When you . . . take[] a situation which could be a simple interview . . . and turn that . . . into a process that [takes] weeks . . . of bookkeeping and meetings and explanation and trainings, what you are simply doing is overburdening people who are trying to do the best.

But the bureaucracy involved in hiring a teacher was nothing like that involved in firing one, as Lombardi explained:

> You need to be prepared to give up two to three years of your life. [It's] very difficult to prove they're incompetent. . . . [F]rom day one you need to document everything. You need to have informal observations, to call the teacher in, to give them a warning that they may get a U rating [i.e., an unsatisfactory rating], to assign someone to work with them, to put them with a professional developer, to provide inter-class visits with them, to do formal observation, to give U-rated observations, and all of these things are grievable if they're file material. [Y]ou give a warning, then you give another warning, then you write a file letter that refers to the warning letter. . . . Every time you give a file letter you must . . . meet with them, you must discuss the issue and you must provide remediation for that issue. You must document every step of the way. Then if you get to the point where you want to U-rate a teacher . . . you have to provide that documentation to everyone. Then you have to go to a hearing, there are always appeals when you U-rate a teacher. If you want to move to dismiss a teacher, there's another issue, then you must . . . get what they call a "tech conference," technical assistance. You have to deal with the court, and so it's very demoralizing.

Part of the problem was that principals were trained to educate students, not litigate, so they were inevitably outmatched, as one of the principals explained in the voice-disguised recorded testimony:

The union has people whose job it is to know exactly how to dissect [a termination case]. That's all they do. . . . It becomes a living nightmare.

I asked Lombardi how long this nightmare lasted "from the time you've identified a poor-performing teacher, to the day the children are no longer under that teacher's supervision?"

LOMBARDI: *Two to three years . . .*
MOSKOWITZ: *[I]f you really put your mind to it, can you deal with the problem in six months?*
LOMBARDI: *No, impossible.*
MOSKOWITZ: *Okay, so, in your experience it's never taken less than two years?*
LOMBARDI: *It can't be done.*

And of course, it was a nightmare for families too. Put yourself for a moment in the shoes of a parent. Imagine you have a second-grader who's been having some difficulties learning to read, perhaps he's dyslexic. You meet with his teacher and she somehow hasn't even noticed your child has a reading problem, so you meet with the principal to express your concern. He tells you this teacher is completely incompetent and that he's in fact in the process of firing her. You heave a sigh of relief since you know your child will soon have a new teacher. But then he explains that the process of firing her will take two to three years. Your intellectually and emotionally vulnerable child will therefore spend an entire critical year of his education in the classroom of a teacher your principal admits is completely incompetent.

It didn't even matter if the case was cut-and-dried:

PRINCIPAL: *The last two teachers that I U-rated were . . . missing between twenty-five and fifty to sixty days of school. [That's] a third of the entire school year.*

MOSKOWITZ: *So you're not even talking about the skill of the teacher. You're talking about a teacher simply not showing up.*
PRINCIPAL: *Right. Which to me is cut-and-dried . . . I don't understand why that should require documents and documents and documents. . . . Two U ratings, they're still there. You cannot get rid of them until they're U-rated three times.*

This was confirmed by a DOE witness:

CHAD VIGNOLA: *You have to . . . provide them with mentoring or other types of supports.*
MOSKOWITZ: *What if the teacher is consistently late? How is the principal supposed to advise the teacher on correcting that problem? Set the alarm clock?*
VIGNOLA: *You would have to put them on notice that being late was unacceptable. You'd have to engage a progressive discipline.*

Moreover, every time a principal U-rated a teacher, the teacher could appeal it at three levels: the superintendent's office, the chancellor's office, and in arbitration.

Even teachers whose poor management skills made them a danger to children were difficult to terminate:

LOMBARDI: *There was a teacher that really couldn't control the class at all . . . The profession was not for her. [One day, an eight-foot-long] window pole goes out the window, almost hits somebody on the fourth floor. . . . I went to the classroom, called the teacher in the hallway, the class is going wild. "Are you missing a window pole?" She looks around, "Yes, I am." I said, "Well, how did it get out the window?" She says, "I apologize, Mr. Lombardi, I was so busy putting out the fire in the garbage can."*

Because it took so long to terminate teachers, Lombardi began taking a different approach. He'd tell a weak teacher that she should consider transferring as he was going to U-rate her but that since U-rated teachers couldn't transfer, he wouldn't U-rate her if she agreed to leave. Lombardi knew this was problematic:

LOMBARDI: *I've been unethical, I'm going to say it in front of the Committee. . . . I made my school better, but they made your system . . . worse, because now they're going to [another] school.*

MOSKOWITZ: *[I]n other words, you gave someone a better rating than they deserved?*

LOMBARDI: *Let's put it this way, during the year I was very critical. At the end of the year we had a meeting of the minds that this might not be the best place for you.*

This practice was so widespread it had a name: "passing the lemon." In fact, we'd heard similar testimony the day before concerning custodians. A principal testified that the only way to get rid of a bad custodian was to rate him highly so he could get transferred to a larger school at which he'd make more money. Rate a custodian poorly and you'd just be stuck with him forever.

After I asked Lombardi questions, it was my colleagues' turn. I understood they didn't share my perspective on the teachers' contract and I figured they might ask questions that got at issues such as protecting teachers from being unfairly terminated. Instead, however, they seemed intent on launching ad hominem attacks on Lombardi:

COUNCIL MEMBER JACKSON: *[A]t what point in time does Tony Lombardi stand up and say I am not going to pass the buck anymore, I'm going to stand up and do what has to be done and give this instructor, this teacher a U rating. . . .*

LOMBARDI: *[I]f I'm going to keep someone three, four years potentially at unsatisfactory ratings, I have thirty students in that class that are affected in day-to-day instruction . . .*

Lombardi had been quite forthright about the conflict between what was in the interest of his students and the system as a whole, but Jackson tried to suggest that he was just being lazy:

COUNCIL MEMBER JACKSON: *So . . . you . . . take the easy route?*
LOMBARDI: *[I] take the route that I can get the best potential out of the students in front of me . . .*

Another colleague of mine continued on this line of attack:

COUNCIL MEMBER RECCHIA: *Okay. So, basically you passed on the problem.*
LOMBARDI: *I admitted to that.*
COUNCIL MEMBER RECCHIA: *. . . so the heck with everybody else, just what's best for Tony Lombardi.*
LOMBARDI: *Well, not specifically for me, for what's best for the school I'm operating, yes . . .*

Here Lombardi was just dealing with the impossible situation in which he'd been put and was having the courage to expose himself to criticism to fix the system, unlike other principals who did exactly the same thing but wouldn't admit to it, but rather than thanking Lombardi for his honesty, my colleagues attacked him as if all of the problems with the entire New York City public school system were due to one principal in Queens who wasn't weeding out bad teachers.

Dan Weisberg, a DOE witness, also testified about how the contract's work rules limited principals' abilities to manage teachers. After I questioned him, my colleague Bill de Blasio took his turn. He claimed that "Oppressive work rules . . . is . . . spin," and that trying to change them was "put[ting] the cart before the horse" because first we needed to provide "competitive salaries" and better "work conditions"—in other words, the UFT's agenda. Weisberg responded that the contract affected teacher morale. He observed that if he was "a new teacher coming in with a lot of idealistic ardor . . . and down the hall I see a teacher who is merely punching the clock . . . at a certain point my ardor is going to start to fade because I see no difference in the way I'm treated versus the way that person who is punching the clock is treated." De Blasio countered dismissively that "every single one of us in our work life has sat in that room where the person next to us is being paid twice as much and was doing half as much work" but that "it's the manager who has to find a way to make the team work nonetheless." "Good managers," he claimed, "find a way to get the most out of people, and make the rules less relevant."

De Blasio's position made no sense to me. Of course, you want a manager to play the hand she's dealt as best as she can, but you also want to give her a better hand to play if possible. Moreover, de Blasio seemed to imagine a world in which every principal is some superstar who can overcome every conceivable obstacle to achieve success. In reality, many principals could barely succeed even in ideal conditions.

Interestingly, ten years after these hearings, Lombardi retired and his school PTA threw him a party at which its president commented: "If a teacher makes a mistake, he's going to call you out on it. . . . He's demanding. He wants success." She also noted "the UFT, they're probably happy he's leaving."[35] She had no idea how truly she spoke.

34

A SHOT ACROSS THE BOW

2013

With de Blasio's election now a fait accompli, I began to contemplate the damage he could do to charter schools as mayor. He'd proposed a moratorium on new co-locations and threatened to revoke those that had already been approved. Success's would plainly be first on the chopping block given his hostile rhetoric. He'd also threatened to make charters pay rent if we received private philanthropy, but we needed those funds to pay the start-up costs for new schools. In fact, if we had to pay market rent, which would be $30 million annually to start, we'd not only be unable to expand, we'd have to slash classroom spending.

Noah didn't wait till it started raining to build his ark, so I began thinking about how we could save ourselves. My fear was that de Blasio would overestimate the strength of his hand. Aside from Success, charter schools had never been terribly active politically because they'd never had to be. De Blasio might therefore assume there would be little political cost to going after us, particularly since he'd been basking in the adulation of supporters who cheered his every attack upon us. In his mind, his ascendance reflected a fundamental ideological shift away from Bloomberg's policies. In reality, he'd beaten weak opponents in a Democratic primary that favored left-leaning candidates. Most New Yorkers did not share his hostility to charter schools. I felt confident that if de Blasio revoked charter co-locations or charged us rent, we could get most of the public on our side. Under the policy of mayoral control for which I'd fought, however, de Blasio held all the cards. Even if we managed to make him regret going after charter schools, he might well stay the course to avoid the humiliation of caving in to pressure. We needed to fire a shot across his bow so he'd realize it was a bad idea to take us on *before* he committed himself.

De Blasio had said that he didn't intend to make every charter school pay rent or revoke every charter school's co-locations so it was clear he intended to use a divide-and-conquer strategy. If other charter leaders let de Blasio treat me as the charter sector's whipping boy, they'd get more merciful treatment when it came time to determine which co-locations would be revoked and how much rent each school would have to pay. In the long run, however, once we were divided, every charter school would be at his mercy. To show de Blasio this strategy wouldn't work, we'd have to demonstrate our determination to stand together. Getting other charter leaders to agree to this, however, was easier said than done. Most were intimidated by power, felt out of their depth in politics, and were reluctant to throw in their lot with a woman the mayor had publicly declared shouldn't be tolerated. My colleagues were political pacifists who believed that every dispute could be resolved through compromise and reason. I subscribe to Frederick Douglass's belief that this isn't always so:

> If there is no struggle, there is no progress. Those who profess to favor freedom, and yet depreciate agitation, are men who want crops without plowing up the ground. They want rain without thunder and lightning. They want the ocean without the awful roar of its many waters.

The obstacle I faced in delivering this message, however, was that some believed I favored conflict because it would advance my political career. I found it strange to be thought of this way. I still saw myself as an educator—a college professor and charter school leader who'd had a brief interlude in politics. It seemed like only yesterday that I was being dismissed as too naive for politics. Now, all of a sudden, I was some modern-day Machiavelli ready to sacrifice children on the altar of my political ambition. Never mind that my support for educational reform had actually hurt my political prospects as demonstrated by my loss in the borough president's race.

But I'd had one stroke of good fortune. I'd mentioned my concerns to Kevin Hall, head of the Charter School Growth Fund, a nonprofit that supported many leading charter schools including Success. Kevin didn't want to see the fund's huge investment in New York charters undermined, so he'd convened a meeting on August 13 with the leaders of the charter schools the fund supported: Morty Ballen of Explore, Brett Peiser of Uncommon Schools, Doug McCurry of Achievement First, and me. He'd urged us to work together to prevent harm to the charter sector, precisely what I was hoping for. Three days after de Blasio won the primary, I reconvened this group on a conference call along with several other charter school leaders. I explained my shot-across-the-bow theory, trying my best to sound like Russell Crowe in the movie *Gladiator* when he says, "Whatever comes out of these gates, we have a better chance of survival if we work together. . . . If we stay together, we survive."

Fortunately, de Blasio had made one miscalculation. When making his proposal that "well-resourced" charter schools pay rent, he probably figured this would affect Success most because our opponents liked portraying us as rich. The reality, however, was that most of the other charter networks raised more money than we did, so his proposal was even more threatening to them. They reluctantly concluded they had to take him on.

We agreed that our first collective action would be a march across the Brooklyn Bridge, ending at city hall. I proposed that in order to make it clear we were willing to stand up to de Blasio, our parents carry signs that said, "What about my Dante?" in reference to de Blasio's son, who had figured prominently in his campaign. I wanted to make the point that while it was great that de Blasio had done well by his son, being mayor was about helping *other* people's children get opportunities. My colleagues, however, were adamantly opposed to referencing Dante and I realized they were right. We were looking to fire a shot *across* de Blasio's bow, not *at* it. Besides, de Blasio would understand the purpose of our march without our taking a jab at him. We agreed instead on the slogans "Charter Schools *Are* Public Schools" and "My Child, My Choice."

With this plan in place, we turned toward organizing our march, which we'd scheduled for October 2. This was just two weeks away and we needed a big turnout; a poorly attended rally would be worse than none at all since the whole point was to show our strength. Luckily, an idealistic young man named Jeremiah Kittredge had recently founded an organization called Families for Excellent Schools to serve as a counterweight to the UFT's parent-organizing efforts. Kittredge agreed to help us organize the march and reach out to charter schools that weren't already part of our coalition.

As October 2 approached, we checked the weather forecasts incessantly since our biggest fear was rain and, sure enough, that's exactly what the weathermen were predicting. When I woke up on the big day, however, it wasn't raining, just cloudy, and just as I arrived at Brooklyn's Cadman Plaza, where the march was to begin, the sun emerged from the clouds. My heart jumped as I saw thousands of parents and children wearing bright green T-shirts bearing our slogans. After listening to inspiring speeches by both parents and teachers, our march began. Everyone's bucket list should include walking over the Brooklyn Bridge with its massive neo-Gothic stone towers, intersecting diagonal and vertical cables, wooden pedestrian walkway, and wonderful view of downtown Manhattan, New York Harbor, and the Statue of Liberty. Added to this inspiring picture on this day was a river of seventeen thousand marchers stretching from one end of the bridge to the other. So many parents came that it took five hours for everyone to cross.

The march was widely covered. The headlines read: "Sea of Parents and Advocates Take to Streets for Charter Schools,"[36] "Parents' Historic March for Charter Schools,"[37] and "Charter School Rally Sends

Message to Bill de Blasio."[38] We'd accomplished what we'd set out to do; now, we'd have to see whether it had the desired effect.

We figured de Blasio's choice of a chancellor might be an early sign of how hostile he'd be as mayor. He ended up plucking a former deputy chancellor named Carmen Fariña out of retirement. I knew her well, as she'd led the schools in my council district and had been principal of my alma mater PS 6. While she wasn't a vociferous charter school opponent, neither was she a reformer, and I doubted she'd stand up to de Blasio if he decided to throw us under the bus. De Blasio's appointments to the Panel on Educational Priorities were also troubling, particularly Norm Fruchter, whose program at the Annenberg Institute had spearheaded the lawsuit against our Cobble Hill school.

Soon after his inauguration, de Blasio announced that he was cutting all capital funding for charter school facilities. While I wasn't happy about this, that money was small potatoes. The real question was what he'd do on co-locations and rent. As I nervously awaited his pronouncements, I began contemplating whether I could strike a deal with de Blasio. While I hadn't appreciated his nasty comments, I was willing to let bygones be bygones if we could agree on something that was in the best interests of the children. I figured I might have to undergo some ritualistic public humiliation so de Blasio could show his more rabid supporters I no longer had "the run of the place," but that was a price I was willing to pay if I could do a deal that was good for our students. Eric said I should be willing to give up co-locations for two of Success's new schools in exchange for preserving the rest of the sector. Our existing schools were a different matter. If de Blasio terminated their co-locations, children we were already educating would be thrown into dysfunctional low-performing district schools. That I couldn't accept. Besides, no matter how much de Blasio despised me, he'd never go that far.

35

SAVE THE 194

2014

On the morning of February 27, 2014, I was at Harlem 1 meeting with teachers and principals when I got a call from Deputy Chancellor Kathleen Grimm. The city, she said, was revoking three of our co-locations: those of two new elementary schools and of our Harlem Central middle school. I gasped in surprise. I couldn't believe the city would really throw kids at one of our existing schools out on the street, so I asked her to repeat what she'd said. She did. I'd heard correctly. What, I asked, would happen to the students at our Harlem Central school? She was silent.

I hung up and broke down in tears. I was stunned. We'd been educating Harlem Central's children since kindergarten. Now, unless we could somehow get this decision reversed, we'd be forced to abandon them.

My mind began racing. I'd have to speak with parents and tell Harlem Central's principal, an idealistic young man named Andrew Malone who had put his heart and soul into creating this school, that all of his effort had been for nothing. It felt like the time I'd lost my first council race. I'd made promises to people, led them to believe in me, and now I was failing them.

I was determined to fight de Blasio but I was by no means confident I could win. Just as I'd feared, he was using a divide-and-conquer strategy. Although he'd made slight adjustments to the co-locations of a few other charter schools, Success's were the only ones he'd revoked. The city did double backflips to justify this result. During the campaign, de Blasio had claimed that Bloomberg's co-locations were problematic because they caused "overcrowd[ing]" and "larger class sizes." He'd now had DOE scrutinize these co-locations for a month and hadn't found a single co-location that should be revoked on these grounds.

However, he still had to make good on his promise not to tolerate me so he came up with brand-new rationales for revoking our co-locations. He revoked the co-locations for our elementary schools, which were slated to share space in high school buildings, on the grounds that it was "more appropriate . . . to co-locate additional high schools" in such buildings because they could share "spaces (such as science labs, shop rooms, and libraries)." This had nothing to do with how co-located schools were actually run. If there were two schools in a building with six science labs, the schools wouldn't share all six rooms but take three each. Our elementary schools could do that just as well as a high school. Indeed, this was exactly what we did at our Upper West and Union Square elementary schools, both of which were co-located with high schools. Moreover, even if there had been merit to de Blasio's rationale, it hardly seemed a sufficiently compelling basis for revoking co-locations that had already been lawfully approved.

The basis for knocking out the co-location for our Harlem Central Middle School was equally flimsy. Because this school was growing too big for its existing co-location, we'd been approved to move into another building that contained one of our existing elementary schools and a special education program, PS 811. To make room for our middle school, our elementary school was going to give up thirteen rooms and PS 811 was going to give up three rooms. As explained in the Educational Impact Statement,[39] DOE's plan was to shrink PS 811's enrollment by thirty students so that PS 811 would have eighty-four students in seventeen rooms, or just over five students per room. The loss of thirty special education seats at PS 811 would be more than made up for by "three new school buildings" opening in Manhattan "that will be able to accommodate . . . approximately 180" special education students.

Most important, PS 811 wouldn't give up its three rooms right away but rather over the course of three years. That meant PS 811 could decrease its enrollment gradually "by placing fewer new students in [PS 811] each year" as existing students graduated. This way, no students would be forced out of PS 811. The administration, however, just lied about this. Chancellor Fariña claimed that "There are kids in that building who would have to leave."[40] De Blasio also said he'd revoked Success's co-location because "We do not want to displace any special ed kids."

But some of the press bought it anyway. The *Daily News* published an opinion piece by my old foe Noah Gotbaum claiming that the co-location would have required "mov[ing] one-third of [PS 811's] autistic and severely emotionally disturbed children out of the building." Similarly, *New York Times* reporter Ginia Bellafante reported that our co-location "would also have displaced a vast number of children with special needs." We wrote to Ms. Bellafante that she was wrong. She responded:

> [The] co-location would force challenged kids who might have been educated at [PS 811] to be educated elsewhere, at other schools in Manhattan, requiring longer or more complicated commutes. That would seem to suit the definition of displacement.

In fact, she had utterly no support for her theory that children would have longer commutes. PS 811 wasn't a zoned school but a specialized school that served students who were bussed in from all over Manhattan. There was no evidence that the three new co-locations DOE was opening up would be any less convenient for prospective students than PS 811; in fact, increasing the total number of locations serving special education students would result in students having shorter commutes on average. Not even DOE had advanced Bellafante's inconvenient commute theory; she'd just made it up out of thin air to justify her misleading article. Incredibly, the *Times* refused to correct the story; you can read it to this day.

Anti-charter ideologue Diane Ravitch also claimed that our co-location required "kicking out students with disabilities."[41] and *Washington Post* blogger Valerie Strauss repeated Ravitch's lie that "special needs student[s] . . . will be evicted" and also wrongly claimed that our "middle school hadn't yet been started," which would certainly have come as news to its students.[42] Thus, she'd managed to completely reverse the facts: she'd claimed that current PS 811 students *would* be affected (wrong) and that current Harlem Central students *wouldn't* (wrong again).

As Mark Twain observed, a lie can travel halfway around the world while truth is putting on its shoes. We tried to give truth a shove by putting out a press release explaining that no students would be forced out of PS 811 and citing the official documents that conclusively proved this, but nobody was interested. "Special education students *not* being harmed" just wasn't a story.

Moreover, just as I'd feared, de Blasio was executing his divide-and-conquer strategy. While Kathleen Grimm was telling me my co-locations were being revoked, de Blasio aides were meeting with six charter school leaders who'd been organized by Jonathan Gyurko, a former UFT operative. The following day, Gyurko emailed Wiley Norvell, de Blasio's deputy press secretary, to give him a "first look" at a statement Gyurko had drafted to see if it "worked" for the administration. This statement, which Gyurko hadn't even shown to other members of his coalition first, supported the administration's decisions, claiming they'd been "principled." This was easy for Gyurko's coalition to say since their co-locations hadn't been revoked. Indeed, the administration had actually extended the co-location for Rich Berlin, whose organization had a board member who had raised substantial funds for de Blasio's mayoral campaign.

De Blasio was playing his hand well. He was isolating Success by only revoking our co-locations, lying about special ed kids being displaced, and deploying Fariña to claim that it was all just a coincidence that the only schools whose co-locations were being revoked were those run by the very person de Blasio had repeatedly lambasted. Thus, we had to tell our story and this is where our level of organization paid off. Our head of communications, Ann Powell, managed to organize a press conference at Harlem Central with both parents and students just hours after de Blasio announced his decision. Reporters immediately recognized that this was a juicy story and covered it sympathetically. "Mayor de Blasio," wrote the *New York Post*, "brought down the hammer Thursday on three charter schools operated by his nemesis Eva Moskowitz, leaving hundreds of kids without classrooms this fall." The *Daily News* wrote, "New York City education officials lowered the boom Thursday on a trio of planned charter schools run by a fiery former councilwoman who is Mayor de Blasio's political adversary."

While media interest was intense, I feared that over time they'd move on to the next thing and that de Blasio would just weather the storm. We had to keep the story alive by showing the real impact of this decision on children. We decided to run an advertisement with pictures of each of the 194 kids at Harlem Central who'd lose their school with the title "Save the 194." We followed up with a TV spot that used the same image. It began:

> These are the 194 faces of Success Academy's public middle school in Harlem. They love their school and all the opportunities it brings. But Mayor Bill de Blasio just announced he is closing their school, taking away their hopes and dreams.

As the narrator said this, the pictures of the students started fading away and the narration continued:

> He's taking away a public school where the fifth-graders have the highest math scores in the entire state . . . and scored twice the city average in reading. Mayor de Blasio, don't take away our children's future. Save our school.

I felt that this ad was particularly strong because it tugged at the heartstrings ("hopes and dreams") while also appealing to logic ("the highest math scores in the state").

Fortunately, the editorial pages of several newspapers rose to our defense. The *New York Post* wrote:

> De Blasio is taking the space away because these kids are learning. That's a huge embarrassment to a mayor and his union allies who spend their time excusing public school failure rather than redressing it. So he's taking it out on the kids. How nasty is that?

In an editorial titled "Bill slams Eva's kids," the *Daily News* wrote:

> Former City Councilwoman Eva Moskowitz is the bête noire of the de Blasio administration. He's aiming at a grown woman—and catching children in the crossfire.

The story soon went national. The *Daily Beast* wrote:

> When does a local education fight become a national bellwether? When it touches a policy lightning rod, scrambles partisan allegiances, and involves political actors who stand in for whole political ideologies. And, sure, it helps when the locale staging the fight is New York City.

As *Slate* observed, de Blasio endured "days of searing press coverage." But there was still the problem that I'd anticipated from the beginning: while we were hurting de Blasio, the goal wasn't to hurt him but to get him to treat charter schools better, and he'd given no signs he was willing to do so. Fortunately, however, we had a plan B.

36

A TALE OF TWO RALLIES

2014

Days before the mayor's eviction of Harlem Central, other charter leaders and I had become deeply troubled by Chancellor Fariña's unwillingness to give the charter sector any reassurances about what the city would do on co-locations and rent, so we'd decided to ask Albany to protect us from de Blasio. While the legislators there didn't love us, we felt that sympathy for our plight might make them more receptive than usual.

The best time to seek Albany's help was when it was negotiating the annual budget, which typically turned into an omnibus legislative package. We therefore decided to hold a rally on March 4. This didn't leave us much time, but, fortunately, Families for Excellent Schools again agreed to plan and organize the rally.

Gyurko's coalition announced they'd be boycotting our rally in part because de Blasio was planning to hold a rally in favor of pre-K funding on the same day as us and they feared we'd upstage him. One coalition member asked, "Is that how you start out the conversation, with a punch in the face?" I was flabbergasted. The mayor had vilified me throughout his campaign and then revoked three of my co-locations and *I* was the one punching *him*? It was like the Woody Allen joke: I'd hit his fist with my face.

Danny Dromm, chair of the city council's Education Committee and a former UFT delegate, called for hearings to investigate Success "just in case there might be corruption" and objected to our plan to close our schools for a day so that our entire community—teachers, students, and parents—could attend the rally. We responded: "We will close our schools for one day to keep the mayor from closing them forever."

But nobody expressed more eloquently what was truly at stake than Harlem Central's principal, Andy Malone, who wrote to his faculty:

> One September afternoon in eighth grade, I hauled my oversize backpack to the public library to begin a research project that would end up changing my life. I was studying the 1961 Freedom Rides. As I read about the rides—a series of nonviolent protests in which hundreds of Americans rode integrated buses through the segregated South—I became completely engrossed. I promised myself that I would dedicate my life to fighting injustice, that I too would be a Freedom Rider.
>
> I came to believe that educational inequality was the civil rights issue of our time. Building great schools for every child in America—this was our Freedom Ride. So after graduation, I set out to teach. Working hard to become a better teacher, and then a better school leader, this was the nature of the fight.
>
> Thursday has changed all of that. Our school has become a political target despite our incredible academic achievements. The hardest part is that it is happening to Laminu and Serea, Vakaba and Amanda, Kwame and Kayla, Jayden and Tiayna, Sharron and Staci, [to our] actors, artists, dancers, basketball players, computer scientists, chess prodigies. In the face of injustice—the lack of logic, the blatant hypocrisy—we see the impact on the kids we know and love.
>
> How do we move forward? The only way is to leverage great teaching as a form of political action, and leverage political action as a defense of great teaching. We need both kinds of Freedom Rides.
>
> Steel yourself for the demands that will come your way: to get every parent to Albany, to invite video cameras into your classroom, to rally and write and speak out. Heed the call to be

> incredible teachers and incredible advocates as we lead a new Freedom Ride toward the right side of history.
>
> We will get there.
>
> *Andy*

On March 4, we held our rally. Eleven thousand parents, students, and teachers from one hundred charter schools statewide participated. Nobody could remember the last time so many citizens had descended upon the capitol to plead their cause. Unbeknownst to me, Governor Cuomo could see them arriving from his office window and was so moved by the sight that he decided at the last minute to speak at our rally. Cuomo was often emotionally reserved in his speeches. He seemed to hold back a piece of himself, to hide the idealism that had led him to choose a life of public service. Perhaps he wished to avoid being perceived as a carbon copy of his father, who was famous for his soaring oratory. This day, however, he spoke with real passion:

> You look so beautiful to me! They say it's cold out here, but I don't feel cold, I feel hot! I feel fired up! We are going to save charter schools and you're making it happen by being here today!
>
> You are here, 11,000 strong. You are braving the cold to stand up for your rights. [T]his is the most important civics lesson you will learn, because this is democracy and this is how you make your voice heard!

Wow! I thought. Calling our rally a "civics lesson" was a hard whack at those who'd criticized us for closing our schools. He continued:

> And we are here today to tell you that we stand with you. You are not alone; we will save charter schools . . .
>
> We spend more money per pupil than any state in the nation; we're number 32 in results. It's not just about putting more money in the public school system, it's trying something new and that's what charter schools are all about . . .

I am committed to ensuring charter schools have the financial capacity, the physical space, and the government support to thrive and to grow.

I was elated. Cuomo had publicly committed himself to solving our facilities problem. Our rally was widely covered in the press, which compared it to de Blasio's:

An overflow crowd of 11,000 charter-school supporters braved Albany's subfreezing weather Tuesday. . . . Meanwhile, de Blasio drew fewer than 1,500 people [leaving] the toasty-warm, Washington Street Armory more than half-empty.

The *Times* published two stories about our rally. One had many pictures of the sea of parents clad in yellow shirts who had descended upon Albany. The other was titled "De Blasio and Operator of Charter School Empire Do Battle" and began as follows:

She was a darling of Mayor Michael R. Bloomberg's administration, given free space to expand her charter schools from a single one in Harlem into a network larger than many New York State school districts. Along the way, her Success Academy empire became a beacon of the country's charter school movement, its seats coveted by thousands of families as chronicled in the film "Waiting for 'Superman.'"

But eight years into her crusade, Eva S. Moskowitz is locked in combat with a new mayor, Bill de Blasio, who repeatedly singled her out on the campaign trail as the embodiment of what he saw was wrong in schooling, and who last week followed his word with deed, canceling plans for three of her schools in New York City while leaving virtually all other charter proposals untouched.

I had mixed feelings about the press's interest in seeing this as a battle between de Blasio and me. I wanted to keep the focus on the critical policy issues at stake rather than me personally, which is why I'd declined to speak at either our Albany rally or our march over the Brooklyn Bridge. On the other hand, the perceived rivalry increased press coverage, which was ultimately a good thing for us.

The next day, a teacher emailed me a picture of one of our students, Sidy Fofana, a boy who played soccer avidly and who, inspired by his cousin's beautiful smile, aspired to become a dentist. The picture showed Sidy, who had attended the rally, sitting next to a structure he'd made in school from wooden blocks. "Sidy built Harlem Central a new middle school building," the teacher explained. "Such a sweet sentiment! If only things were this simple." It made me realize how incomprehensible it must be for children that anyone would want to take away their school.

At a press conference, Fariña was asked what would happen now that our co-location had been reversed and she responded, "They're on their own now." It was one of those answers that, like Marie Antoinette's "Let them eat cake," was perfectly wrong. The press jumped on her comment as heartless. Two days later, she apologized and said that the city was now "looking for additional space that might accommodate" our school. While I liked the sound of that, it wasn't clear to me whether she was promising she'd actually find us space or just give it the old college try.

I was invited to appear on the political talk show *Morning Joe* and then on March 10, de Blasio was invited on the show to give his response. After playing the clip in which de Blasio said I shouldn't be tolerated, Joe Scarborough asked, "What don't you like about Eva Moskowitz? That seemed awfully personal." De Blasio denied it, but when pressed to explain why PS 811 couldn't lose seats given that three other special-needs schools were opening, he punted by saying the decision had been DOE's and he had to "trust" they'd made it correctly. De Blasio was dodging this question, not because he was too busy to learn the facts—after all, he was appearing on a national news show precisely to discuss this very issue—but rather because there was no good answer.

He did, however, say that his administration was going to find "an alternative for Success Academy." This seemed to confirm Fariña's comment although we still had no details. Where would it be? Would it be permanent? Would he make us pay rent? Bizarrely, we had to learn everything from TV and newspapers just like everybody else because the administration wouldn't speak with us.

Given this uncertainty, we decided to initiate claims before the State Education Department and the federal courts. As any reader of this book now knows all too well, state law required detailed notices and multiple hearings to change a building's use. Moreover, de Blasio had promised to go beyond those legal requirements and do additional preliminary engagement of "stakeholders." Instead, he'd simply announced his decision to revoke our co-locations as a fait accompli without even putting in a single call to us, much less holding the necessary hearings. His hypocrisy would have been comical if the fates of real children weren't at stake.

We followed up on our rally by arranging for dozens of meetings between our parents and their representatives in Albany. These meetings put a human face on our schools and allowed us to educate legislators and their staff about charter schools and parents' dissatisfactions with the district schools. I met with the head of the assembly's Education Committee, Catherine Nolan. She wasn't a big fan of charter schools but, unlike some of our opponents, she had common sense and a good heart. She seemed genuinely surprised by de Blasio's actions. From our contact with legislators, we also came to understand that de Blasio had alienated many elected officials in Albany with his pre-K rally because it had felt to them like an attempt to force their hand rather than plead his cause.

On March 13, the senate majority proposed legislation that would reinstate the co-locations de Blasio had revoked, prohibit him from revoking any more or charging us rent, and require the city to provide new and expanding charter schools with either facilities or the funds to rent them. Predictably, Gyurko drafted a statement opposing the legislation which said that his coalition "would prefer a solution that includes the direct involvement of the affected communities, families, and other stakeholders." This infuriated me. De Blasio was refusing to speak directly with the "families and other stakeholders"—namely us—which was exactly why we needed Albany's help.

Shortly thereafter, Cuomo expressed his views in a radio interview:

> If a charter school is not given a location and is not given funding to find a location, then in essence the charter school is out of business. We're not going to be in a situation where charter schools stop, okay? Not if I have anything to do with it.

On March 16, the *New York Times* finally weighed in:

> [T]he three co-locations that were initially rejected involved schools run by Mr. de Blasio's longtime political rival, Eva Moskowitz. . . . Ms. Moskowitz, who has a flair for the dramatic, staged a rally in Albany, hit the airwaves and has threatened a civil rights lawsuit. Both sides should stop escalating the conflict, which helps no one.

I found it amusing to hear that my protests reflected a "flair for the dramatic," as if I'd gotten my knickers in a twist over some small slight. As for de Blasio and me being longtime political rivals, that was flat out wrong. I'd never run against de Blasio for office or said one negative word about him until after he'd won the primary, at which point he'd been attacking me publicly for months. Finally, the claim that escalating the conflict was "helping no one" was even more silly; it was helping kids! It had gotten de Blasio to promise he'd find space for Harlem Central and I was hoping it would get him to find space for our elementary schools as well.

Later that day, I learned that several of my coalition partners had agreed to meet with de Blasio. I was upset that they hadn't consulted me since we'd agreed at the beginning to formulate our strategy together. This was part of de Blasio's divide-and-conquer strategy and it worried me. To date, de Blasio had only succeeded in getting Gyurko's group on board; far more damage would be done if de Blasio succeeded in chipping away at the coalition I'd formed. I got them to delay the meeting so we could come up with specific demands to present, such as committing to approving a certain number of colocations, so that de Blasio couldn't pull the wool over their eyes by talking in general terms.

On March 18, a poll came out showing that only 38 percent of voters approved of de Blasio's handling of the public schools. A subsequent poll showed voters particularly disapproved of his handling of charter schools. Billy Easton of AQE, a union-funded education advocacy group, sent a memo to de Blasio recommending that he "quickly give a major education speech," which de Blasio did on March 23.

In writing about this speech, the *Times* observed that de Blasio had sought to strike "a conciliatory tone, acknowledging missteps and emphasizing common ground." That sounded like a swell idea if it meant he'd no longer denounce me as public enemy number one, but I was more interested in learning what he planned to do about our three schools, so I read on. Quoting a theologian, de Blasio said, "The noise of these shallow waters prevents us from listening to the sounds out of the depth." He wanted, he said, "to create a fullness, a totality, a completeness." I didn't know what this meant but it sure didn't sound like "Sorry I took your buildings, you can have 'em back."

After this excursion into existentialism, de Blasio got down to business. "The answer," he said, "is not to save a few of our children only[,] to find an escape route that some can follow and others can't [but] to fix the entire system." This was a false choice. Sure, no single school, whether district or charter or parochial, can alone fix the system. But create *enough* good schools and they will. Every good school, of whatever kind, contributes to the solution.

De Blasio did not acknowledge that evicting our schools had been a mistake and accepted fault on only one point: he hadn't "measure[d] up when it came to explaining [his] decisions" and on that score, he promised "to right the ship now." In other words, he was going to do a better job of telling us why he'd been right all along. Once you cut through his "shallow waters" mumbo jumbo, he wasn't giving an inch.

Predictably, his speech was praised by the Gyurko coalition, which de Blasio had continued to court by personally meeting with them and even promising them space not only for their schools but also their pre-K programs as well. All sorts of goodies would be made available just so long as they continued to side with de Blasio against me. Deputy Mayor Richard Buery stroked Rich Berlin by emailing him "you have a big fan in the mayor." "We are on his side big time," Berlin responded, "and will help however we can." It astonished me that a fellow educator could so enthusiastically support our eviction.

Now we waited for Albany to act and they soon did. On the afternoon of March 28, just as the meeting between de Blasio and the other charter school leaders was about to take place, Cuomo and the legislature announced they'd reached an agreement on a new law that would prohibit the city from charging rent or revoking any existing co-locations and would require the city to provide new or expanding charter schools with either facilities or a rent subsidy. It was a huge victory. "The law," observed the *Wall Street Journal*, "represents a reversal of fortune for Mr. de Blasio." "It turns de Blasio's campaign promise on its head," observed another commentator. "Instead of Eva Moskowitz paying him rent, he'll be paying her rent."[43]

So we celebrated. The legislature had come to our rescue and saved our schools. Or so we thought.

37

TRAGIC THINGS HAPPEN

2014

On March 29, the day after the legislative agreement was announced, we got to see the legislative language. No charter school co-location, it said, could be "revoked, overturned, or withdrawn, nor shall any such [co-location] that has not been . . . withdrawn . . . as of the effective date of this act fail to be implemented." I emailed Emily Kim and Eric: "This language seems great, no?" Not really, they responded. The city, the law said, had to implement co-locations *except those that had already been withdrawn,* and de Blasio had arguably *already* withdrawn ours.

Surely, I thought, this hadn't been the legislature's intention given that de Blasio's treatment of our schools was the principal reason for the law. It would be like the Greeks leaving Troy without Helen. We tried to get the law fixed in what was called the "cleanup bill," a law that was passed to fix late-night drafting mistakes made in the annual budget law, but nobody wanted to reopen this can of worms. So we'd worked 24/7 for months on end, brought thousands of parents to Albany, and won an incredible victory by getting historic legislation passed—only to find ourselves back at square one and running out of time.

We came up with an idea for a compromise to propose to the city. Our middle school had been slated to move because we needed more room as we added seventh and eighth grades, but the school's existing grades—fifth and sixth—could fit in the building in which the school was currently located. Thus, if we left those grades where they were, and only put the seventh and eighth grades in the PS 811 building, we'd need only a few rooms which we could get from the Success elementary school that was already there. Though splitting our school wouldn't be ideal for us, it would completely resolve the city's concerns since PS 811 wouldn't have to give up one single room. On April 2, Emily wrote a letter to the city proposing this solution. The city quickly rejected it, however, and wouldn't explain why. If I had any doubts whatsoever that the city's actions were political, this removed them.

We sought to keep the pressure on by reactivating our media campaign, asking Cuomo to intercede, and amending our claims before SED to argue that the new law protected our co-locations (although we knew this was debatable). On April 3, we had a conference call with the city, which was represented by Deputy Mayor Tony Shorris, Zachary Carter (the city's top lawyer), and a half dozen other staff members, a surprisingly large and high-powered contingent. Emily and Eric represented Success. Shorris offered us a former parochial school building in West Harlem for our middle school but asked us to delay opening our two elementary schools for a year. We'd consider the parochial school, Emily said, but not delaying our elementary schools. Shorris then asked us to postpone our lottery, claiming that it would be irresponsible to go forward with them since the city didn't have sites for our elementary schools. When Emily refused, Shorris claimed the city had the legal authority to prevent us. This made Eric livid. Not only did the city lack such authority, it was obvious that Shorris's real concern was that it would be harder for the city to wriggle out of finding us space once we'd selected students. Eric observed that we'd unfortunately become quite accustomed to running lotteries in the face of uncertainty, as the de Blasio administration was just the latest in a string of opponents trying to prevent us from opening schools. We did agree, however, to tell SED it could delay deciding on our petition for ten days to give the city time to find us space for our schools.

I feared Shorris was just trying to run out the clock on us, to string us along until it was too late for us to open and the press had lost interest, so we'd therefore insisted that the city propose new school buildings for us by April 7. When the city missed this deadline, we let them know we planned to hold a press conference the following day. Cuomo somehow got wind of this, and offered to oversee our negotiations with the city if we didn't go forward with our press conference. I couldn't turn him down, but I worried he'd pressure me to reach a compromise that would be bad for our schools and the families we served. I wasn't against compromise where there were legitimate competing interests, but I saw no reason to sacrifice the needs of children just to serve de Blasio's political agenda. Moreover, I felt that this was as good a time as any to bring these issues to a head. De Blasio wanted to show that there was a new sheriff in town and that he could slow down charter school growth. I wanted to show the opposite: that he'd underestimated the public's support for charter schools and that Bloomberg's policy of providing co-locations should continue. De Blasio was now trying to beat a strategic retreat from the fight he'd started because he was losing, which was precisely why I wanted to finish it. However, to avoid giving Cuomo the impression that this was some type of personal vendetta against de Blasio, which it wasn't, I'd have to remain calm, which would be challenging since Shorris had a smartest-guy-in-the-room swagger that rubbed me the wrong way.

We set a meeting for April 11 at 2 p.m. in Cuomo's Manhattan office. The participants were me; Shorris; Cuomo; and Larry Schwartz, Cuomo's chief of staff. Shorris offered us two elementary school sites: one in Jamaica, Queens, and one in midtown Manhattan. The Manhattan site was only available for a year, however, because the city had already agreed to sell it to a developer. That, I said, was a nonstarter; I couldn't admit students to a school with no idea where it would be located a year later. Moreover, midtown would be a tough commute for the families, who were from downtown, the school's original location. Shorris said dismissively that this wasn't a big deal, that he lived in Brooklyn and managed to get his kids to a school in Manhattan. This comment infuriated me. Shorris had pulled down more than a million dollars a year at his last job, so he could afford taxis and babysitters. Getting kids to a distant school was much harder for poor families. Shorris's comment was symptomatic of a bigger problem. People at the top level of government were out of touch. Sure, some sent their children to public schools, and never failed to pat themselves on the back for doing so, but only the best ones. De Blasio's children, for example, had attended schools in Park Slope, one of the city's chicest neighborhoods, and then Beacon and Brooklyn Tech, two of the city's most selective high schools. Similarly, Shorris sent his child to the most selective public school in the entire city. These schools bore utterly no resemblance to the dismal schools the children who'd won our lottery would have to attend if our schools couldn't open.

Seeing my frustration rise, Cuomo asked Shorris and Schwartz to step outside. I feared he was now going to insist I accept a bad compromise. Instead, speaking in a frank but friendly manner, he said he just wanted to make sure I understood that if I didn't compromise, I might lose and walk away with nothing. I responded that I did understand this but was not willing to give up on opening our schools. What I didn't say was that I regarded defeat as preferable to surrender. A compromise with de Blasio would be an implicit endorsement of the outcome. If de Blasio was going to kill these schools, to take away these educational opportunities parents wanted for their children, I wanted the world to know that he'd done it. This was the best way to ensure that de Blasio didn't do to other charter schools what he'd done to us. Then, at least, our loss would serve some purpose.

Cuomo called Schwartz and Shorris back in and told us to keep working toward a solution. Shorris demanded a media cease-fire and I agreed to give him another week to come up with more concrete proposals that specified not only what buildings they were offering but how long we'd have them, who'd fix them up, and whether the city would provide security and maintenance as it did for district schools. A week later, the city hadn't clarified any of this nor come up with a permanent location for our Manhattan elementary school. I extended my deadline to April 22, but when the city failed yet again to meet it, I went through with my press conference. "I thought this was settled," said one of our parents. "The mayor went on television and said he'd find space. He's stated repeatedly that he's a public school parent—surely he must know how we must feel, not knowing where our kids will be next year."

Reading about all of this drama, you might think that this was what I did with most of my time, but I still had schools to run and crises arose daily. For example, I got safety alerts whenever a serious safety issue arose at one of our schools such as nearby gunfire, child abuse, or student injuries, and in April alone, I received twenty-nine of them. I also had to make sure instruction remained excellent; winning a battle to get buildings from de Blasio would be a Pyrrhic victory if it so distracted us from instruction that our schools declined in quality.

One day, April 24, can suffice to reveal how crazy this period was. We had three major events scheduled for the following week: the state math tests; a move to new headquarters; and our annual gala, the biggest fund-raising event of the year. On April 24, I woke up at 5 a.m. to find an email stating that Cuomo wanted to meet with me and Shorris at 1:30 p.m. An hour later, Richard Seigler called to tell me that Sidy Fofana, the boy who'd built a pretend building for Harlem Central from blocks, had died from falling off the roof of a forty-three-story building. I was horrified. How could such a thing happen? A nine-year-old boy gone? I thought of his poor parents and of his teachers. After calming myself, I called Keri Hoyt and Vanessa Bangser, the principal of Bronx 2, to get their help. By 8:35 a.m., Keri had formulated a detailed plan that involved dozens of people, including arranging for the psychologists from all of our schools to speak with Sidy's teachers and his classmates and their families. Meanwhile, Vanessa and I both went to Harlem 3 to help out. By 9:30 a.m., we'd notified all of Harlem 3's parents about this tragedy and arranged meetings to tell Sidy's classmates and their parents about his death.

Next, I raced downtown to sign off on the renovations at our new headquarters and then to my meeting with Shorris and Cuomo. Shorris complained that finding a school facility in downtown Manhattan impossible. This was exactly what I'd feared: that he'd simply throw up his hands at a certain point and declare that the city couldn't solve the problem it had created.

About a month earlier, Eric had told me that a building in northern Manhattan was being vacated by Mother Cabrini, a parochial school. Since it was at the other end of Manhattan, it wasn't a good replacement for our downtown school. However, I now accepted that I would have to give up on opening a school downtown, but I figured that if I could get the city to rent the Mother Cabrini building for us, at least we could meet the needs of another community. Thus, much to everyone's surprise, I made this proposal and Shorris agreed to look into it. I added, however, that I wasn't granting any more extensions. SED was set to rule the following day, April 25, and if we didn't reach a deal by then, I'd tell SED to go ahead and decide our legal claims, let the chips fall where they may. Everyone wanted to avoid that, so we agreed the parties would meet the following day beginning at 9 a.m. and wouldn't leave "until all terms on all three proposed sites are agreed upon."

But my day wasn't done yet. A teacher had witnessed a parent push her child angrily and then flee the building, so we reported the incident to the authorities and arranged for the child's grandmother to pick her up. I was also concerned that students at our other schools would learn of Sidy's death and might have questions about it, so I wrote the following to our teachers:

> We wanted to make sure you had resources to help you tomorrow should scholars or families ask you any questions about what happened.
>
> Attached you will find a document that should be helpful. In addition, here are answers to the frequent questions we got at Harlem 3 today.
>
> **How did he fall?** We don't know.

Why did this happen? Tragic things happen and we don't always know why they do. It can be very upsetting, but that's why we're all here to support one another.

What happens when you die? Not all families believe the same thing, so it's really something to talk with your family about.

Throughout the day, I'd been furiously emailing every moment I could—in elevators, in cabs, in the bathroom, and while walking, which I did by trailing a few feet behind my assistant like a duck following its mother. In the midst of Sidy's death, and all that we were doing to help our children through that tragedy, I was dealing with other matters such as our high school math curriculum, making sure we had copies of the state tests we were about to give, reviewing our practice test data for the upcoming state math tests, overseeing our schools' participation in a charter school study, negotiating with a university that was helping train our teachers, planning our gala, improving our computer science instruction, monitoring our investment in our database systems, and approving the payment of forty-two purchase orders totaling more than $1 million. At 10:49 p.m. that night, I sent my final email in which I forwarded a request for information from one of Cuomo's aides.

That was my April 24.

The following day, I attended Sidy's funeral along with most of Harlem 3's teachers. As Sidy's family was Muslim, we wore headscarves and bowed down on our knees for most of the funeral, which was conducted in Arabic. There is nothing more painful than attending a child's funeral. All of that potential, that joy Sidy brought to school every day, had evaporated in an instant.

While I mourned, our facilities team negotiated. It consisted of Emily, Pete Cymrot (who worked with Emily), and Kris Cheung (our head of facilities). The city had a far larger contingent, including representatives from DOE, the School Construction Authority, and the law department. A top Cuomo aide, Ian Rosenblum, oversaw the negotiations.

Emily had drafted a term sheet covering the minutest of details concerning renovations, maintenance, and security for each of the buildings. The city initially balked at many of the terms, but Emily is a ferocious negotiator. When the city objected to building an AstroTurf field for our Queens school, Emily showed them pictures of the football stadium to which we'd have had access at our original site. Hour after hour, the negotiations went on and when nighttime came, the city's negotiators began leaving one by one. Those who remained, eager to join their colleagues, became increasingly pliant. At 1:31 a.m., Emily wrote me that the deal was done. I felt an enormous sense of relief. For six months, we'd been fighting this battle. Finally, we were done, having achieved both a good result for our students and a law that would help the entire charter sector.

However, to say all's well that ends well would be disingenuous. It was an enormous waste of time, energy, and, most important, taxpayer money. The city soon projected it would spend more than $5 million per year on our three school buildings, a completely unnecessary expense.[44] It was also an enormous distraction for de Blasio's administration. He'd complained that Bloomberg had focused too much on charter schools and said he wanted to shift the attention back to district schools. That's what he should have done. However, instead of focusing his efforts on making the district schools better, he'd tried to hamstring charter schools, which put them front and center. Diane Ravitch—no fan of mine—would later ask: "How did a [network of schools] that serves a tiny portion of New York's students manage to hijack the education reforms of a new mayor with a huge popular mandate?"[45] The answer is that we didn't; de Blasio hijacked his own agenda.

Some people feel I'd wanted to pick a fight with de Blasio. Nothing could be further from the truth. I wanted to focus on education. I don't enjoy conflict and I'm naturally a pretty cowardly person. I can't stand to watch violent movies, I dread getting shots at the doctor, I jump five feet in the air if someone sneaks up on me, and I worry about everything. But I wasn't going to let de Blasio take away our schools. The philosopher Lao Tzu said "loving someone deeply gives you courage." So does loving a cause.

Sidy's death continued to weigh upon the Harlem 3 community and upon me personally. Several days after his funeral, I wrote to my faculty: "Schooling is about giving life to kids. Schooling is about getting kids to make the most out of life. Death is the opposite of everything we do." Sidy's best friend was particularly distraught and said, "It should have been me." This boy disappeared one day and his mother found him on the roof of their building. We gave him counseling and made sure to keep a close eye on him.

To give the Harlem 3 community closure, we held a memorial service for Sidy. As students were there, we focused more on celebrating Sidy's life than mourning his death. The ceremony began with remembrances by Sidy's teachers and a reading of a letter from the dean of NYU's School of Dentistry honoring Sidy's ambition to become a dentist. Richard Seigler presented college scholarships in Sidy's honor to his four siblings and to one of his classmates. At the suggestion of one of Sidy's teachers, we ended the ceremony with a soccer game by Sidy's teammates to celebrate his passion for this sport.

The mystery of Sidy's death, of how he'd ended up on the roof and why he'd fallen off, was never answered. A small consolation was knowing that in August, Sidy's younger brother Salif would be entering Harlem 3's kindergarten. It was a reminder that amid death and tragedy, children bring hope and renewal.

38

RELEASING ATOMIC SECRETS TO THE RUSSIANS

2003

By the second day of my contract hearings, even the balconies were full, which was unheard of for a city council committee hearing. Word had gotten around that Randi Weingarten would be testifying and this was the showdown for which everybody had been waiting. Even journalists who weren't covering the hearings were there just to watch. Given the intense media and political interest—the council chambers had the electric atmosphere of an arena before a heavyweight boxing championship—I worried about remaining calm and composed.

The success of the hearings would turn in large measure on how well I handled Weingarten, which worried me since she was a tough customer. Before joining the UFT, she'd been a big-firm lawyer and could be quite aggressive. I learned this the hard way when I'd appeared with her on television and had been unable to get a word in edgewise. I feared she'd now filibuster me by giving long speeches and refusing to answer my questions. Ordinarily, a chair could exercise a certain amount of control over committee proceedings, but if I tried to rein Weingarten in, my colleagues might take her side, claiming I wasn't giving her a chance to speak. The path I had to walk was a narrow one: giving Weingarten a fair chance to present her views but not giving her so much leeway that she hijacked the proceedings.

Weingarten immediately went on offense, attacking Lombardi, and bragging about the fact that she'd gotten the other principal to cancel his appearance because "he knew we were going to show [his] anti-union animus," and claiming I was "demoniz[ing]" teachers. As for the summaries we'd provided:

> The inaccuracies and the misconceptions . . . are too numerous for me to rebut here. We'll do that in a separate document, but you can be sure, Eva, that our separate document will . . . be printed by a union shop.

No such summary was ever released.

Weingarten went on interminably. Her opening remarks alone ran more than eight thousand words, longer than the entirety of Lombardi's testimony, including his opening statement. Among other things, Weingarten claimed it took just 65.5 days on average to terminate a teacher. It sounded very objective, particularly with that ".5," but it was quite misleading because it referred only to the length of the dismissal hearings themselves. Lombardi and other principals had testified that they couldn't even start the dismissal proceedings until they'd spent at least two years building a record: giving feedback to the teacher, U-rating the teacher multiple years in a row, and fighting the teacher's grievances every step of the way. I asked Weingarten about the estimates the principals had given me:

> MOSKOWITZ: *So they're way off?*
> WEINGARTEN: *Whatever [DOE's] internal processes are, that is a management responsibility . . .*
> MOSKOWITZ: *There's nothing that requires them to go through an extended grievance process?*
> WEINGARTEN: *No, there is not.*
> MOSKOWITZ: *And if they don't document, will they succeed?*
> WEINGARTEN: Eva, *in any hearing that somebody goes through, somebody has to get up and say why they want somebody terminated.*

This was particularly disingenuous because, as Weingarten knew full well, the UFT had a whole staff of attorneys whose sole mission in life was to make it impossible to terminate a teacher without creating an extensive record that included multiple U ratings.

I also asked Weingarten questions about seniority, but getting her to answer them was virtually impossible. She'd go on long tangents or attacks or quibble with terminology. I finally expressed my frustration:

MOSKOWITZ: *These are simple straightforward questions. Is it the case under the UFT transfer plan that a principal could have to accept a candidate sight unseen; is that correct, or incorrect?*

WEINGARTEN: *Again, there is no, and I don't mean to be a pest, but there is no UFT transfer plan. There is a plan that was negotiated starting in the mid-sixties between the Department of Education and the UFT.*

MOSKOWITZ: *Could you answer my question? Is it the case that a principal under any circumstance could have to take a candidate without ever having interviewed them; yes or no?*

WEINGARTEN: *Yes.*

Getting that "yes" took five minutes. Being a trained litigator, Weingarten knew that I had a limited amount of time and that the more obstructive she was, the less ground I could cover.

Weingarten claimed that only 10 percent of vacancies had been filled through seniority because most schools were using the alternative "school-based option" for hiring teachers. However, even this approach required that candidates with the most seniority be hired unless "a less experienced applicant . . . possesses extraordinary qualifications." Moreover, in order to get this very limited flexibility, the principal had to give up control over hiring to a personnel committee "[t]he majority of [whose] members shall be teachers selected by the UFT chapter."

But rather than get into those complexities, I focused on a much simpler point, which was whether seniority rights were a good thing:

> MOSKOWITZ: *[Y]our point is that the seniority transfer option is only exercised by 10 percent, but if my kid has one of those teachers, I don't care because it's my child . . .*
> WEINGARTEN: *Is the chair saying to me that you believe that anybody who exercises the seniority transfer is therefore a bad teacher?*
> MOSKOWITZ: *Of course not. You're implying that because it's only 10 percent it may be okay, and I'm suggesting if it is a bad thing, and it affects 10 percent of the vacancies, we have a problem. So, let's get to the larger issue, which is whether it's a bad thing or not. And I'm asking.* I'm asking.
> WEINGARTEN: *First off, I am not implying that seniority transfers are a bad thing or a good thing. I am telling you—you've asked me, and I guess this is something where you clearly do not believe, and if you want to get to seniority, let's talk about it, you clearly do not believe that seniority or experience . . .*
> MOSKOWITZ: *I'm not going to have you tell me what I believe and don't believe. What I'd like to understand is your view of it.*

At this point, Weingarten launched into yet another one of her filibusters, this one totaling 419 words, whose essence was the following:

> WEINGARTEN: *[I]f you read the contract, what it does say is that . . . qualifications trump everything. [I]f people are equally qualified, then the most experienced person tends to get a job.*

This sleight of hand showed Weingarten's lawyering skills. The only thing the contract said about "qualifications" was that a properly licensed teacher was "qualified." For an English position, an applicant licensed in English was more qualified than an applicant licensed in math. However, if you had one hundred candidates who were licensed to teach English, they were considered equally qualified, so you had to hire the most senior one even if another candidate had won the Nobel Prize in Literature.

I next asked Weingarten whether she objected to the Education Committee examining the teachers' contract. She said "Eva, from the first moment you raised this with me, I never had an objection to you looking at the contract." Given the UFT's efforts to get the hearings canceled, this was certainly news to me! I pressed her further, however, and she explained that she objected to the timing of the hearings because negotiations were ongoing. In other words, she didn't want the hearings to be held at a time when they could actually have an impact.

After I finished, my colleagues pitched Weingarten softball questions that allowed her to repeat many of the points she'd already made. She'd also brought along a teacher whose purpose became clear when one of my colleagues started asking her questions:

COUNCIL MEMBER RECCHIA: *Talking about Principal Lombardi, now, you worked for him; is that correct?*

ROSEMARIE PARKER: *And I still do.*

RECCHIA: *Is it a fact today, does he have two classes of special ed in his fourth . . .*

MS. PARKER: *He has no special ed classes in the fourth grade this year . . .*

RECCHIA: *And isn't it a fact that the reason why his fourth-grade reading scores went up is because he got rid of the special ed students in the fourth grade?*

MS. PARKER: *[H]e has declined the amount of special ed classes in the school so their scores cannot impact the overall scores of the school.*

This line of inquiry was completely irrelevant, and had been orchestrated simply to embarrass Lombardi. Given that the UFT was willing to launch such an obviously ad hominem attack on Lombardi in a room crowded with journalists, I feared what they'd do to him when those journalists weren't around. The next day, he'd have to go back to school and work with this woman who had just publicly attacked him. This was why it had been so hard to get any other principals to testify.

Council Member de Blasio used part of his time to rebut my contention that witnesses were being intimidated although he'd just witnessed it happening. He, however, soon made it clear that he supported only "whistle-blower[s] in the classic sense," which apparently didn't include those who criticized labor contracts.

The third day of hearings would echo an issue I'd raised in my very first campaign, principal tenure. Tenure for a management position struck me as particularly problematic. Someone who asks to be put in a position with greater compensation and power should also accept greater accountability. Moreover, the tenure rules ultimately made it harder for a principal to succeed because they applied to assistant principals as well. Thus, while DOE had recently given principals more control over their budgets—whether to buy new computers or new textbooks, and whether to hire a volleyball coach or a bandleader—when it came to their most critical decisions, the selection of the management team that would help them carry out their vision, they were hamstrung, as Daniel Weisberg, DOE's executive director for labor policy, explained:

> [L]ifetime tenure rules . . . make it next to impossible to discharge an assistant principal. . . . [A]lso . . . a principal who recruits a promising new assistant principal and invests the time to train him or her, can have the assistant principal bumped out of his or her position at any time by a more senior assistant principal displaced from another school.

The contract also made it harder to turn around failing schools:

> Principals . . . may inherit substandard assistant principals who do not share their goals or priorities. . . . If principals were able

> to freely choose their assistant principals, we [could] install entire turnaround teams of principals and assistant principals in chronically failing schools.

Former principal Bernard Gassaway testified about the problems he'd faced:

> I inherited nine assistant principals . . . [who had] contributed to the downfall of that particular school. . . . [I]t took me . . . three to four years, actually, before I was able to shape my team. . . . While I'm battling my team, so-called team members, I'm not giving the attention . . . that I need to give to . . . instruction.

I asked why the DOE hadn't changed the contract:

> MOSKOWITZ: *Did the department want to give principals [the] ability to select their assistant principals?*
> WEISBERG: *Yes.*
> MOSKOWITZ: *And the [administrators' union] opposed that?*
> WEISBERG: *They filed an improper practice charge with [the Public Employee Relations Board] challenging that reform, yes.*

Why would the administrators' union oppose giving its own members the ability to select their assistant principals? Because it represented both principals and assistant principals and there were more of the latter than the former.

The final witness that day was the administrators' union president, Jill Levy, who said we had no business examining the contracts because we lacked "jurisdiction . . . in the collective bargaining process." Given that the council was responsible for approving a school budget of around $10 billion, most of which was spent on labor, the idea that we shouldn't worry our pretty little heads about how this money was being spent struck me as an astonishingly narrow view of our responsibilities to children and taxpayers.

The following day, we heard from our final witness, Chancellor Klein. He spoke thoughtfully and eloquently about the impact of the labor contracts upon the school system and his determination to make changes; I'll spare you the details, however, as it covered much of the same ground as the witnesses who'd preceded him. After he testified, I brought the hearings to a close.

While those who attended the Education Committee's hearings learned how the city's labor contracts affected the public schools, press coverage was required to reach a broader audience. Fortunately, we got it in spades. Ironically, it was the unions' opposition to the hearings that ensured their success. We'd take on a topic that had been "all but taboo until the Education Committee's hearings,"[46] and the result, observed *The Economist*, was "four days of drama."[47] A Bloomberg aide commented "You'd think she was releasing atomic secrets to the Russians."[48] A *New York Observer* columnist wrote:

> City Council hearings generally pass with little or no notice in the media. Not these. . . . Ms. Moskowitz's confrontation with Ms. Weingarten was particularly memorable. As political theater, it doesn't get much more dramatic than this.[49]

Moreover, while the drama was the draw, the press also got into the substance by describing the "arcane work rules" the contracts contained such as those against custodians painting above ten feet, changing ballasts, vacuuming rugs, or replacing more than seventy-five floor tiles per month; against teachers monitoring corridors or supervising lunchrooms; against paying bonuses to principals or higher salaries to teachers in hard-to-fill positions such as science; against transferring assistant principals or selecting teachers based on suitability and merit rather than seniority; and against terminating teachers without wading through a paralyzing thicket of grievance procedures.

As for the Bloomberg administration, while it had initially viewed the hearings as a "distraction," that changed:

With each passing day—as the public learns about custodians who won't vacuum and bad teachers who cannot be fired—many in the administration are starting to see the upside. "What we are seeing," said one Bloomberg aide, "are all the contractual things that make it very difficult to improve the way children are taught."[50]

I was pleased that the hearings had contributed to public understanding of these contracts. What I hadn't appreciated, however, was the impact these hearings would come to have on my own views.

39

EVA MOSKOWITZ AND I ARE SAYING THE SAME THING NOW

2014

In 2005, Weingarten had started the UFT charter school to use "real, quantifiable student achievement" to disprove the "misguided and simplistic notion that the union contract is an impediment to success."[51] The UFT school, she'd said, would be the "perfect environment for the UFT to demonstrate that its educational priorities work."[52] Since the UFT and I had started charter schools at around the same time and with markedly different views on education, it was not lost on anyone that the fates of our respective schools could have a broader meaning. Indeed, one justification for charter schools is permitting educators to experiment with different approaches so you can see what works. Now, nearly a decade later, the results were in.

At the nine Success schools with students old enough to take the state tests in 2014, 94 percent of the students passed the math test. Our schools outperformed not only New York City (35 percent) and New York state (36 percent) but also the city's most affluent school district, District 2 (66 percent), and one of the state's most affluent suburbs, Scarsdale (68 percent). Of the state's top ten schools in math, four were Success schools. In science, *all* of our students—*every* student at *every* school—passed the exam. Ninety-nine percent got the top score of 4.

In English, 64 percent of our students passed, outperforming New York City (29 percent), New York state (31 percent), and the city's most affluent school district, District 2 (56 percent). The passage rate for our English language learners was eight times that of the city's district schools (41 percent versus 3 percent) and was also higher than even the *non*-English-language learners in these schools.

At Success Academy Bronx 2, located in the poorest congressional district in the state, 99 percent of our students passed the state math test. Out of New York state's 3,528 schools, Bronx 2's poor children of color bested nearly every other school in the state—including schools in wealthy suburbs and gifted and talented schools that selected their students. First place went to Success Academy Upper West, where 100 percent of our students scored proficient. At Bed-Stuy 1, where 95 percent of the students were minority, 81 percent scored proficient in English and 98 percent in math, with 80 percent receiving a 4. At Harlem Central, which de Blasio had sought to close, 96 percent of our students scored proficient in math.

Some people try to explain away our results by saying that we serve fewer poor and special needs kids than the nearby district schools. Consider Bronx 2. Eighty-eight percent of our students there were Title I (meaning poor); at the district school in the same building, PS 55, 96 percent were Title I (8 percent more). Fourteen percent of our students had learning disabilities; at PS 55, 15 percent did (1 percent more). But while the differences between our students and PS 55's were miniscule, the differences in results were huge: ninety-nine percent of our students were proficient in math compared to 15 percent at PS 55; 70 percent of our students were proficient in English compared to 7 percent at PS 55. Clearly, an 8 percent difference in poverty and a 1 percent difference in special needs doesn't explain an 84 percent difference in math proficiency and a 63 percent difference in English proficiency.

And by the way, even our Title 1 students were 97 percent proficient in math and 77 percent proficient in English. Our special needs students? One hundred percent of them were proficient in math. They beat their classmates. Go figure.

As for the UFT's schools, 2 percent of its eighth-graders passed in math (compared with 97 percent of Success's eighth-graders) and 11 percent passed in English (compared with 94 percent at Success). Soon after getting these results, the UFT announced that it was closing its elementary and middle schools.

I give Weingarten credit for putting her educational theories to the test but, when those theories failed, she sought to distance herself from her experiment. She claimed through a spokesperson that her "involvement with the school[] had ended years ago . . . shortly after she resigned as [the UFT's] president in 2009" to become president of the national teachers' union.[53] In fact, she chaired the school's board until at least September 30, 2010,[54] a school year in which only 13 percent of the school's eighth-graders scored proficient in math and only 22 percent in English.[55] Thus, the school was already sinking by the time she abandoned ship. Ironically, when I'd founded Success, she had demanded to know whether, if I failed, I'd "blame others . . . [o]r . . . take responsibility for what goes wrong?"[56] Yet, not only did she fail to take responsibility, she refused to reconsider the theory that her experiment was designed to test that the UFT contract wasn't "an impediment to success."

Other members of the UFT's board were Zakiyah Ansari, the spokesperson for AQE, one of the UFT front groups, and ACORN founder Bertha Lewis, who had criticized "privatization of public education"[57] but for some reason considered the UFT's charter school exempt from this charge. Since 2010, the UFT Charter School had been run by Shelia Evans-Tranumn, the SED regulator who'd killed our developmental kindergarten program. Under her leadership, the school's performance went from bad to worse. Incredibly, the school even failed to comply with the regulatory requirements that Ms. Evans-Tranumn had enforced in her prior career, including performing criminal background checks on staff, having a Finance Committee, and complying with the Individuals with Disabilities Education Act and Open Meetings Law. While it may seem mean-spirited to make this observation, I do so to underscore a point. While regulations are crucial in ensuring that schools don't discriminate, abuse students, or cheat on tests, giving regulators the power to dictate a school's instructional approach—its curriculum, its selection of teachers, its policies for disciplining students—simply transfers authority away from the people on the ground, who are in the best position to make these decisions, and are responsible for the school's results, to distant bureaucrats who have neither any proven competence in running schools nor any accountability for the results of their decisions.

Comparing our results and those of the UFT's school illustrates an important point about standardized tests. Suppose such tests didn't exist and you, a reader of this book, were trying to evaluate my claim that Success's schools are better than the UFT's school. I could wax on eloquently about how all of our children are critical thinkers and creative problem solvers, are self-confident and self-actualized, write like F. Scott Fitzgerald and dance like Fred Astaire, and have wonderful portfolios of their work that I so wish I could append to this book so you could see them for yourselves, but alas that's impossible (although perhaps I could include a couple of "representative" essays). The problem is that the UFT's principal could make the same claims. How would you know who was telling the truth? You'd have to visit these schools and carefully compare student work at each school for children of the same age across multiple grades and multiple subjects. This just isn't practical for you, or for parents, voters, or even superintendents of school districts. That is one of the reasons standardized tests are so important. Even if they don't give you the whole picture, they are a practical and objective way to tell more or less how much a student is learning or how well a school is doing. The fact that 97 percent of our eighth-graders passed the math test, compared to 2 percent of the eighth-graders at the UFT charter school, may not tell you everything you'd like to know about these schools but it tells you something.

Given our strong results, we applied in 2014 to open fourteen more charters, an unprecedented number, and were approved to do so, which led to an amusing *Daily News* editorial titled "Bill, Hug Eva" which encouraged de Blasio to work with me. Alas, I received no hugs, but de Blasio did exhibit a wariness toward me like that of a dog toward a porcupine it has once attempted to bite. As for the Gyurko coalition, while the administration had courted them like Casanova, once the battle with me was over, the administration's ardor quickly cooled as *Capital New York* reported:

> Some [coalition members] said they came to believe that the mayor simply does not like charter schools. . . . [T]he soured relationship with City Hall ultimately made the charter sector more homogenous in many of its policy goals and in its frustration with de Blasio . . .
>
> "If their goal was to unite the sector, mission accomplished," said one coalition member who asked not to be identified. "Eva Moskowitz and I are saying the same thing now."

40

EVA MOSKOWITZ IS GOING TO SAY SOMETHING TOMORROW

2015

A school should prepare children to be productive members of society but it should also help them appreciate the life of the mind and enjoy culture and art for its own sake. Knowing how important travel had been for me, I decided we should take some of our students on a trip to Greece and Turkey. Among those selected was Sydney McLeod, but I feared her health would prevent her from going. Her doctors discovered that, in addition to sickle cell anemia, she suffered from CVID, an immune system deficiency that can result in chronic infections. Just a few months before this trip, Sydney was admitted to the hospital with a bone infection. The doctors gave her a blood transfusion and performed surgery that required drilling into her bones. When she was released from the hospital, she needed a wheelchair, so we arranged for a daily car service so she could continue her studies.

Gradually, however, her health improved, and by the summer of 2015, she was well enough to travel. She wrote of this trip:

> In Athens, we explored the Acropolis and the ancient agora of Athens, and walked where some of our greatest ancestors—Socrates, Plato, Pericles, and Euripides among them—had gone before us. Some of the stone slabs beneath our feet were the very ones that these "greats" of western civilization walked on. The Greek Parthenon stuck with us the most—what an incredible sight. We were

able to see how massive and beautiful the structure was from the great white columns to the wonderful inscriptions. And have we mentioned the sacred precinct of Delphi and the eternal city of Ephesus? These sites helped us look at Greece's history and made it come alive!

As I read this, I thought of the bicycling trips Eric and I had taken on which we'd seen these very same sights. I was so happy that Sydney and her classmates were also being given the opportunity to experience them.

Fortunately, our board members have helped expose our students to many cultural and professional opportunities. Daniel Loeb has brought Garry Kasparov, Salman Rushdie, and George Stephanopoulos to visit our schools. Tali Farhadian Weinstein, a former federal prosecutor, took our students to federal court where they met with prosecutors and a judge. In the fall of 2015, we were able to open up a pre-K program because we'd gotten the legislature to authorize charter schools to run pre-K. Given the demise of our DK program, I was glad we'd finally be able to serve younger children, but then de Blasio found a way to prevent us. The law specifically stated that "all . . . monitoring, programmatic review, and operational requirements . . . shall be the responsibility of" our authorizer, meaning SUNY. This meant the city couldn't regulate our program, which was critical since de Blasio had never met a regulation he didn't like. The city insisted, however, that we sign a contract that dictated every aspect of our program down to the minute and, when we refused, wouldn't pay us, so we had to cancel our program. We sued and on June 8, 2017, five appellate judges ruled unanimously in our favor reasoning that "all" actually does mean "all." As I write this, however, the city has refused to say whether it will actually comply with the court's ruling.

While I kept my focus on schooling, the press was increasingly viewing me as a potential mayoral candidate due to my high-profile battle with de Blasio. My detractors claimed that my every action was in service of a Machiavellian plot to become mayor. It simply wasn't true but this perception was undermining my credibility as a charter school advocate. I therefore figured I should decide whether I really wanted to run for mayor in the next election cycle and, if not, take myself out of the running.

The circumstances for running were propitious. De Blasio had antagonized many New Yorkers by making some impolitic remarks concerning the police; traveling to Iowa to engage in national politics rather than focusing on his job; arriving late at events, including a memorial for victims of a plane crash; and, of course, trying to evict our schools. Moreover, as the *Times* observed, I was "uniquely capable of reassembling the political coalition that coalesced behind Michael R. Bloomberg" and "a natural choice for a hodgepodge of communities frustrated by Mr. de Blasio, including white voters in Manhattan who have soured on the mayor, business leaders who have long viewed Mr. de Blasio with hostility and a diverse set of charter-school parents across the city."

I believed I could do a good job of managing the city, particularly with the experience I'd gotten running Success. I was also troubled by de Blasio's leadership. While his commitment to progressive politics was no doubt sincere, it often expressed itself in class-warfare rhetoric that was imprudent and dangerous. While the issue of income inequality is a serious one, it must be approached delicately in an age when hedge fund managers can work from anywhere in the world with an Internet connection. Scare away the rich and you'll kill the goose that laid the golden egg. In addition, so much of de Blasio's outlook seemed driven by this oppositional view of politics: rather than promote "A Vision of One City," he attacked "A Tale of Two Cities"; rather than articulating his own vision of education, he attacked Bloomberg's. Finally, he had an ends-justify-the-means attitude toward ethics that I'd seen at work on the council and that manifested itself in several scandals in his administration.

I was sorely tempted to take on de Blasio, and many people were encouraging me to do so and offering financial support. I was reluctant, however, to abandon Success. Not only were we changing the lives of ten thousand children directly, we were increasingly having a broader impact both by sharing our best practices with other schools and advocating for charter schools and reform. In theory, I'd have even more power to do good as mayor, but in practice the unions and the political establishment would likely fight me tooth and nail on the changes I'd want to make. I decided not to run.

Now, I had a practical problem. I wanted to announce my decision but to do so in person because I knew there would be a lot of speculation about my motives and I wanted to answer reporters' questions directly. I needed to set up a press conference but I didn't want to reveal in advance what I'd be announcing or the journalists would publish stories before I had a chance to explain myself. Therefore, on October 7, I announced that the following day I was going to hold a press conference regarding my "political plans." To my amazement, my email provoked a firestorm of speculation. I was intentionally "build[ing] suspense," one reporter tweeted.[58] Another claimed the whole thing was a hoax. A third was so impatient that he published an article with the bizarre headline "Eva Moskowitz Is Going to Say Something Tomorrow."[59]

De Blasio tried to diminish attendance at my announcement by offering reporters "a rare off-topic question and answer session" at his own competing press conference.[60] It didn't work. Dozens of reporters showed up to my press conference and, after a brief statement, I answered every question they had. I think that in the end, even those reporters who didn't agree with my views at least respected me for believing in something and were disappointed they'd be deprived of a ringside seat at the de Blasio–Moskowitz title challenge.

One of my motives for my announcement was to diminish the press's interest in me and Success so I could focus on schooling. Rather than escaping the limelight, however, we got more coverage over the next sixth months than ever before—none of it good.

41

THE DOMINANT NARRATIVE ABOUT CHARTER SCHOOLS

2015–2016

In nearly two decades in the spotlight, I've had considerable dealings with the press, and by and large they do their job well. In the field of education, there is particularly good work being done by reporters who have decided to devote their career to writing about education. Many of these journalists don't work for major papers but rather focus on publishing magazine articles or books or write online either on their own blogs or for sites such as *Chalkbeat* and the *Hechinger Report*. While there is also some good work done in major newspapers, they sometimes assign relatively junior reporters a tour of duty on the education beat before promoting them to cover matters considered more important. That practice produces decidedly mixed results, particularly if the reporter in question doesn't have any prior experience in the field of education.

Good journalism is absolutely critical, and that is why our country has generous legal protections for the press. Even when journalists are flat-out wrong and their errors cause real damage, they usually can't be held liable unless they had "reckless disregard for the truth." This is how it should be, but some journalists abuse this privilege. The next two chapters are about journalists whose bias led them to get things wrong with very serious consequences for Success, our wrongly impugned educators, and the charter school movement.

In September 2015, I learned that a public television correspondent by the name of John Merrow was going to let a former Success student's mother (whom I'll call Jane Doe) talk about our treatment of her son on the air but not let us respond:

John,

[I'm told] you intend to allow Jane Doe to speak about our treatment of her son but refuse to allow us to tell our side of the story. That is unfair.

Eva

Dear Eva,

Our story is . . . not about Jane and her son. We would not air unsubstantiated accusations . . .

John

John,

I don't understand. If Jane and her son are not part of the story, why is she going to be on the air? If she is making accusations about the treatment of her son, we'd like to know what they are and respond.

Eva

Dear Eva,

Because Jane was unwilling to release her son's records, we were of course unwilling to allow her to openly criticize the school. Her role in the piece is limited and should not be a cause for concern on your part.

Our piece also emphasizes—"celebrates" might be a more appropriate verb—your network's focus on science and the arts, its remarkable academic success, and its widespread popularity.

Respectfully,

John

On October 12, the segment aired. It didn't "celebrate" our "academic success" or our "focus on science and the arts." Instead, it criticized us for allegedly "suspend[ing] very young children over and over" for trivial misconduct. Jane Doe and her son were his sole on-the-record sources for these allegations, despite Merrow's assurances that they wouldn't be allowed to "openly criticize" our school. Jane Doe said her son had been suspended for "meltdowns," "cry[ing]," and "outbursts," and he said he'd gotten in trouble for "wearing red shoes" and not "keep[ing his] shirt tucked in." In fact, our records showed that her son (whom I'll call John Doe) had been suspended for innumerable violent incidents including throwing a classmate against a wall and threatening "to use [the boy's] head like a soccer ball"; lifting his desk above his head and attempting to throw it at other students; kicking, scratching, and punching a teacher; throwing a stapler; punching a teacher and continuing to do so even after the teacher said, "You're hurting me," requiring safety guards to be called; and choking a teacher. On one occasion when John misbehaved, his mother said, "I'm gonna beat your ass" and told her son that she intended to have the police lock him out. On a museum field trip, John began "screaming" and then "kicking the walls and doors," so a security guard restrained him. John's teacher called Ms. Doe, hoping that she could help calm him down by speaking to him over the phone. This is the teacher's contemporaneous account of what happened next:

> Before I could speak, Mom started screaming on the phone, "Let go of my son!" John hit the phone out of my hand and onto the floor, ran away from the museum worker, and took my phone. Mom kept calling my phone and I told John he could answer, knowing it was Mom, but he refused. [Another teacher] tried to calm John down, but he then went screaming and running around kicking people, slamming doors and screaming. [The other teacher] had her cell phone out and John took hers and threw it across the room with

mine and both broke. Two museum workers had to restrain John as he kicked and screamed. A museum worker then asked us to leave the building.

Unfortunately, John behaved this way often. Frankly, it was only by applying a very lenient standard that John was suspended only eight times over nearly three years. Had we suspended him for minor infractions, as Merrow claimed, he'd have been suspended daily.

I wrote PBS demanding a retraction. As I explained in my letter, not only was Merrow's account regarding John Doe a complete fiction, so too was his claim that Success had a high rate of student attrition due to suspensions:

> New York City's Independent Budget Office and WNYC public radio have found attrition in district schools [is] 13 percent annually.[61] Success's is 10 percent according to data obtained by WNYC.[62] Thus, Success has a *lower* attrition rate than most district schools [or] charter schools (10.8 percent).[63]

PBS issued a "clarification" conceding we "should have been given a chance to respond" to Jane Doe's claims but stood by Merrow's report because other anonymous parents had allegedly made similar accusations. PBS seemed utterly untroubled by the fact that, with the claims of Merrow's sole on-the-record source disproven, the entire report now rested on anonymous sources whose claims weren't and couldn't be fact-checked. As for attrition, PBS claimed Merrow's analysis was based on "internal documents" but refused to share with us this secret stash of data that supposedly proved that both WNYC and the Independent Budget Office had gotten it all wrong. (Months later, Merrow got into a fight with WNYC that ended with his deleting portions of a blog entry critiquing WNYC's methodology.)[64]

Although journalists are loath to criticize one another, Merrow's violations of journalistic ethics were too egregious to overlook. One journalist pronounced Merrow's work "shoddy."[65] Another marveled that a seasoned reporter would ignore such a "basic journalistic requirement" as allowing us to respond.[66] The *Washington Post* called Merrow's emails to me "damning" and rejected PBS's claim that Merrow's report could stand independently of the debunked claims regarding John Doe: "To any objective observer, Merrow's story most certainly was about" John Doe and PBS's claim to the contrary was "what you call Ex Post Facto Story Redefinition." Even PBS's own ombudsman felt compelled to speak out:

> [A]ll the critics are anonymous except two, the mother and her son. And she won't release her son's records, which probably should have been another red flag signaling that perhaps this youngster did do some pretty bad things. . . . There is, for me, just too much in this presentation that depends on anonymous "Eva Moskowitz's critics," and "other parents told us" and "but our sources . . . charge."

Anonymous sources can sometimes be used responsibly, but this wasn't one of those instances. First, Merrow's anonymous sources were addressing a topic on which people find it hard to be objective: their children's flaws. That's particularly true of parents whose children have behavioral challenges. Thus, it's hardly surprising that a half dozen of the ten thousand or so families we serve would claim their children had been unfairly disciplined. Second, this was an instance in which anonymity fundamentally undermined Merrow's ability to fact-check his story. That's not always true. Suppose an anonymous source claims a company has fired its CFO because he was a whistleblower. The reporter could then ask the company why it fired its CFO, see if it holds water, and, at a minimum, report the company's version. Here, however, Merrow's unwillingness to reveal the names of the children in question prevented us from pulling up the relevant records and giving him our side of the story regarding these students.

Working with anonymous sources requires that journalists maintain a healthy level of skepticism but Merrow just assumed that any parent who'd criticized Success was being truthful. He did for the same reason he lied to me about his story, refused to let us respond to Jane Doe, and created his bogus secret statistical analysis: he didn't believe young children should ever be suspended. Of course, Merrow was entitled to his own opinion about whether we should suspend young children, but not to his own facts about why we do.

Another journalist named Dana Goldstein commented regarding Merrow's piece that "even little kids are capable of violent behavior that impacts other children's ability to learn" and that the problems this presents "are some of the toughest in education." Critics of tough discipline policies like to put their head in the sand and focus on only the interests of the child being disciplined, forgetting that the needs of other students are at issue as well. In the museum incident, for example, the entire class had to leave the museum, thus losing this educational opportunity. A violent child can also hurt other children. A parent whose child is in a class with such a student expects the school to do something to make sure the classroom is safe. Reasonable minds can't differ about how a school should balance the needs of a student who is having difficulty behaving with the needs of that student's classmates for a safe and productive learning environment, and a serious examination of that dilemma would be a worthwhile public service. Instead, however, Merrow preferred to pretend this challenge didn't exist by falsely claiming that we were suspending children for minor infractions. In peddling this fiction, Merrow not only denigrated our educators, he also failed his viewers.

Far more damage, however, was inflicted by a series of articles by *New York Times* reporter Kate Taylor. Two weeks before Merrow's report came out, she let us know she was writing about an incident at our school in Fort Greene, Brooklyn, where I'd appointed a new principal, Candido Brown, ten months earlier. Like many of our educators, Candido's commitment stemmed from personal experience, as he'd explained to another reporter soon after his appointment:

> I attended schools where we didn't get homework. I was afraid to go to school. I didn't do anything all day. I didn't learn to read until later.
>
> My mom was a drug addict. I was taken care of by my sister, faked many days of sickness because I hated school. I was afraid I would be picked on or beat up.
>
> My work today is to ensure my children come to a school that's safe, that's rigorous, that's warm, that's engaging, that's fun.

Candido had his work cut out for him because our Fort Greene school had a lot of problems. I'd been trying to fix those problems and in fact had appointed Candido in the hope that he could turn things around. Only two weeks after he took charge, I got a report suggesting he was succeeding:

> Parents have responded really well to [Candido's] presence. They feel they have access to him and that he is being far more transparent than his predecessors. We are all happy with the changes he's making. The tone of the school is far more calm and controlled.

Four days later, however, a troubling email was brought to my attention. Under the heading "got to go," it listed sixteen students whose behavioral problems Candido apparently felt were too severe for the school to handle. Now, it's true that some students have learning or behavioral challenges that are so severe that, both for their own benefit and those of other students, they need to attend District 75 schools like PS 811, the school at issue in our fight with de Blasio. Nearly twenty-four thousand children, about 2 percent of New York City public school students, attend such schools. None of our principals, however, had ever claimed to have sixteen such students. Candido's list had to include students whose behavior, while no doubt challenging, could nonetheless be addressed in a conventional school setting.

I suspected Candido had given short shrift to his obligation to help challenging students overcome their behavioral problems because he wanted to ensure his students wouldn't suffer from the same type of unruly and unsafe school environment he'd experienced as a child. Nonetheless, he violated our policies, so I immediately took action. I emailed my managing director of schools, Deanna Durrett, to get her help:

> Not only did I react strongly to language in the email but I know Candido to be stubborn. He needs a strong wake-up call.

I put Success's Michele Vespi, our managing director of talent, in charge of organizing the meetings that would deliver this wake-up call and she soon reported back to me that Deanna had already had "an extensive conversation" with Candido and that he was "confirmed to come to the Network tomorrow" to speak with both Michele herself and Emily Kim. Thus, we took this so seriously that we were having three of our top people meet with Candido: our general counsel, our managing director of schools, and our managing director of talent.

Deanna reported back to me afterward that the meetings had gone well and that Candido now understood "how problematic his words and approach were." Candido himself wrote:

> I apologize for this incident and will be more mindful moving forward. I am dedicated to all our children and want to ensure each one receives the education he/she deserves.

I was pleased with this response but I nonetheless appointed Danique Loving, one of our most experienced and capable principals, to keep a close eye on Candido.

Happily, the message seemed to get through. We require teachers to keep detailed records and these records demonstrate that our teachers made valiant efforts to help the students on Candido's list and that, far from trying to push out these students, teachers had positive communications with them. Here are some excerpts from these records:

1/8/2015: "Called mom to let her know that this week has been one of [student]'s best in the whole year."

1/25/2015: Email to parent: "This week went really well. I'll continue to practice the strategies you have shared. We want him to be successful, so if there are any other strategies you recommend, please let me know! I appreciate all your help in communicating with me and with [student] so he can feel successful! Let me know if there's anything else I can do!"

2/2/2015: Email to parent: "I can totally tell you have been working with him, he was very enthusiastic today and his work and focus was much improved. He shared with the class today. I have attached a picture of him sharing." Mom's reply email: "You just brought tears to my eyes!!!! Thank you so much. I will continue my efforts!!"

3/9/2015: "Called to let [parent] know that though [student] had fourteen corrections, he was still putting forth his best effort all day with no crawling under furniture. Mom was very pleased to hear about this."

3/9/2015: Email to parent: "[Student] had a really good day. Both Ms. B. and I were working on ensuring that we used calm, neutral tones when talking to him so as to not upset or anger him when giving corrections. I was informed we are going to put [student] in [another classroom]. Ms. B. and I both will miss him very much. We have poured a lot of time and love into him. I wish him all the best and will be checking in on him." Mom's reply email: "Thank you. I appreciate all of the hard work and effort that both you and Ms. B. have put into educating [student]."

4/15/2015: Teacher called mom to tell her "[student] earned special pizza lunch with [Candido] Brown [for] following directions, and doing his work. Mom was happy."

4/15/2015: Email to parent: "Here is a picture of [student] on his second time-in [i.e., a chance for a child to play], which was ten

minutes long because he earned four stars! He said his playdough is a T-Rex named [Candido] Brown! The work you did with him has helped him tremendously today. Thank you so much for your hard work and support with him. We could not be more grateful."

Plainly, our teachers were trying to help these kids adjust to Success, not trying to get them to leave, although their behavior was often quite challenging, as our records reflect:

"[Student] hit the scholar behind him."

"[Student] jumped out of his seat, grabbed a pencil, and stabbed another scholar in the finger."

"Another scholar was raising his hand to ask a question and [student] swatted at his hand twice. The boy yelled stop and [student] went to hit again but I stopped him."

"Sent [student] home early for two-handedly pushing teacher into garbage pail."

"[Student] turned around and punched another scholar."

"[Student] took off his shoe and threw it at me and it hit my shoulder."

"[Student] 1) had the folder over his shoulder, looking ready to swing it down on another scholar; 2) after intercepting that hit by running across the room and standing between [student] and the other scholar, he threw the folder down on the table; 3) when asked to step outside the room, he yelled at me, said no, kicked over the garbage bin, and continued yelling; 4) I was finally able to talk and walk him over to the door frame. When I asked him to breathe (some of his calming strategies), he hit my rib cage with a closed fist; 5) when other teachers came to help, he did not follow directions."

Given that we had some challenging students and a well-meaning but relatively green principal, I felt we'd handled this situation as well as could be expected. Candido had written his list on December 9 and by December 12, just three days later, we'd reprimanded him, he'd acknowledged his error, and the school was now sincerely trying to help these kids with their behavioral issues.

Ten months later, Kate Taylor told us someone had given her Candido's list and she intended to write about it. We told her that Candido's actions didn't reflect our policies and gave her virtually all of the information above, but it made no difference. On October 29, the *Times* published Taylor's article, which bore the title "At a Success Academy Charter School, Singling Out Pupils Who Have 'Got to Go.'" At every turn, she'd crafted her article to make it appear that Candido's list reflected our policies. For example, only after Taylor built up the case against us with nine lengthy paragraphs littered with the claims of anonymous sources (sound familiar?) and of a parent who said her child had been "treated unfairly" did Taylor finally get around to letting the reader know that Candido had been reprimanded and, even then, made it sound very unconvincing:

> In a written response to questions, Success Academy's spokeswoman, Ann Powell, said that the "Got to Go" list was a mistake and that the network quickly got wind of it and reprimanded Mr. Brown.

Taylor left out all of the details about the reprimand: the fact that we'd brought Candido into the Network, that we'd done so the day after the list had been brought to our attention, that the reprimand had been delivered by three of Success's top people, and that Candido had apologized for his conduct. To the reader, it therefore sounded like a typical unsubstantiated media flak denial.

Taylor also completely ignored the information we'd given her about our teachers' efforts to help students. We'd even put her in touch with several parents on Candido's list whose children were treated so well that they had no idea anyone at Success had ever thought their child should attend a different school, but Taylor managed to make even that sound nefarious:

> [W]hen a reporter asked if she knew that her son had been included last year on the "Got to Go" list, Ms. Cooper said she did not.
>
> "I'm a little upset about that," she said after a minute. "They could have let me know he was on a list that he 'had to go.'"

Thus, Taylor made it sound like our success in making Ms. Cooper feel her child was welcome was somehow dishonest.

Of course, just like Merrow, Taylor bolstered her story with anonymous sources who'd supposedly told her that "some administrators singled out children they would like to see leave." This sounds damning, but it's unclear what it means. As noted above, 2 percent of district school students have sufficiently serious intellectual or mental health issues that they need to attend special programs. We also get some such students. In other cases, Success just isn't the best fit for a child. Just as some college students do better at big universities while others are more suited to a small liberal arts college, Success isn't ideal for every child. If we think a child would do better in a different school, whether it's a specialized program or just a school with a different approach, we'll tell a parent that, as we should. Yet Taylor characterized this as "singling out children we would like to see leave."

What bothered me most was Taylor's unwillingness to put this incident in context. For seven months, we'd been getting reports that she was calling former employees to investigate the claims the unions had long made that we "counseled out" students. After all that digging, she'd come up with one case where, literally, our least experienced principal—he'd been on the job exactly twenty days—had made a mistake, had been reprimanded for it within three days, and had thereafter made sincere efforts to mend his ways. And moreover, those reprimands had been delivered not as a PR response to a press report, but ten months before Taylor had called us. To me, this proved that counseling out kids *wasn't* our policy. According to Taylor, however, Candido's list was proof that our "accomplishments are due, in part, to a practice of weeding out weak or difficult students."

Let's examine that claim. Our Fort Greene school didn't have students in testing grades that year, but our nearby school in Bed-Stuy did and 99 percent of its third-graders, 75 out of 76 students, got a 4 on the state math test. In District 14, where that school is located, 9 percent of students got 4s. If the difference in our scores was due to the fact that we simply counseled out weaker students and kept the strong test takers, then in order to admit 75 strong test takers, we'd have to admit 833 students in total (since 9 percent of 833 is 75) and then counsel out 758 of them. If we were counseling out even a fraction of that number of students, our attrition rates would be *dramatically* higher than those of district schools, but they are actually *lower*. And yes, we explained all this to Taylor and, yes, she ignored it.

In 2016, WNYC public radio did what Taylor should have. It set out "to get beyond anecdotal reports" and look at the data on "whether, on the aggregate, tough discipline and even suspensions at charters would correlate with parents choosing to pull their kids out of these schools."[67] In particular, they "took a close look at Success Academy, which has been in the news lately."[68] (No kidding.) Their team included veteran education reporter Beth Fertig and "data gurus John Keefe, and Jenny Ye" who liked to "look at numbers dispassionately."[69] Just prying loose the data from DOE took a year. When they completed their analysis, they made public both their results and, unlike Merrow, the underlying data. They found no link between attrition and charter schools' tougher discipline policies. In fact, at Candido's school, attrition was 21 percent *lower* than the schools in the surrounding district. Network-wide, our attrition rate was 43 percent *lower*.

Yet, despite the *Times*'s purported concern about our attrition, it neither reported on this analysis nor did its own. Rather, as Alexander Russo of the *Washington Monthly* observed, WNYC's analysis "was largely ignored" because it "contradict[ed] the dominant narrative about charter schools." WNYC's solid and dispassionate work went unrecognized while Taylor's misleading and sensationalist reporting ricocheted across the Internet. Sadly, there seems to be little correlation between the quality of a journalist's work and the platform they're given.

And alas, Kate Taylor wasn't done with us yet.

HOW MUCH DO PARENTS KNOW OF WHAT GOES ON IN THEIR CHILDREN'S CLASSROOMS?

2016

On January 11, 2016, Taylor showed us a cell phone video which, unbeknownst to us, an assistant teacher had taken a year earlier. It showed one of our teachers ripping up a student's paper and saying to her angrily "You're confusing everybody. I'm very upset and very disappointed." We immediately suspended this teacher, Charlotte Dial, and interviewed her colleagues, her students, and their parents to find out if she'd acted similarly on other occasions. We found she was universally regarded as a capable and loving, albeit demanding, teacher who did not ordinarily act as she had on this video. Moreover, she acknowledged her behavior had been inappropriate and told us she'd apologized to her class at the time. By interviewing the student who'd been reprimanded and her mother, we learned that the student didn't recall the incident and hadn't mentioned it to her mother at the time. While that doesn't excuse Charlotte's conduct, I was nonetheless glad to learn this student hadn't been traumatized by this incident.

We decided not to terminate Charlotte but rather to suspend her and retrain her on interacting with students positively and respectfully.

As for Taylor, we suggested she interview the parents of Charlotte's students to put this incident in context. Here are excerpts from that interview:

PARENT: *[Ms. Dial] was very nurturing . . . we absolutely loved her.*
PARENT: *Very supportive, motivating, encouraging. I was overjoyed this year when I found out my son also had her.*
PARENT: *My child knew she loved him.*
PARENT: *I've been [a teacher] for twenty-five years so I've been in dozens, if not hundreds, of schools. No other school [has] as much warmth and nurturing as there is here.*
PARENT: *This is just an unfortunate minute and a half out of an otherwise amazing experience.*
KATE TAYLOR: *And have you or any of you ever seen anything like that?*
PARENTS: *No.*
KATE TAYLOR: *And how often are you in your child's classroom?*
PARENT: *Every day.*
PARENT: *[Ms. Dial] is in the job she belongs in. She loves it and she loves children.*
PARENT: *I drove her a bunch of times back and forth to the Bronx for our soccer games, and she's so devoted to closing the achievement gap. It's just what her life is. She's been doing it for seven years.*
PARENT: *Those seventy seconds could have been handled better, but that is not who she is, and that definitely is not who we are as a community and as a school.*
PARENT: *We can't judge her or the whole school based on that. If you walk into any classroom now you'll see a very loving and caring environment.*
PARENT: *This is an amazing school, and you hear that from 999 out of 1,000 families here. Was that one teacher over the line for sixty seconds? Yeah. Do I want that teacher removed? Not at all.*
PARENT: *We are not going to send our kids to a school where they're being emotionally abused. Parents are very aware of whether our kids are happy or not in school and a lot of us spend a lot of time in school. I am also a former education reporter.*
PARENT: *It seems like there's a whole lot of good stories here and they don't seem to make their way into the* Times.

TAYLOR: *Talking to all of you is exactly to get this context. . . . [I] certainly recognize the fact that you are all here, and some of you have come from other schools and are extremely happy here, that's obviously very significant. It says a lot. And I recognize that.*
PARENT: *And that will actually be printed? I mean, it's a side that you're going to share?*
TAYLOR: *I'm not talking to you so that I can* not *use it.*

In fact, this was exactly what Taylor was doing. In the lengthy article she soon wrote, Taylor didn't quote a single positive parent comment. Instead, she briefly quoted one parent saying that he'd be troubled "if you tell me that happens every single day," but that "everyone is telling me about all the amazing things that she does all the other days." It sounded like this parent, unlike the others, wasn't sure how Charlotte ordinarily behaved ("if you tell me . . .") and that he was relying on others, presumably Success itself, to tell him what was really going on ("everyone is telling me . . ."). Taylor didn't mention that the other parents were confident they knew what was happening in Charlotte's classroom because they visited it regularly and spoke to their children often, or that many were sophisticated educational consumers including a former education reporter, a teacher, and parents who'd taken their kids out of other schools. Nor did Taylor mention any of the comments about Charlotte being loving, nurturing, encouraging, and supportive.

But much more important, Taylor falsely claimed that Success encouraged teachers to act the way Charlotte had in the video, relying on a couple of former teachers whose demonstrably inaccurate claims she failed to fact-check. I'm going to use pseudonyms for these teachers because I'm sympathetic to the difficulty they had in viewing the circumstances of their departure objectively. One, whom I'll call Jane, claimed we'd encouraged her to "belittle" children and had resigned because she "no longer wanted to be part of an organization where adults could so easily demean children." That, however, was flatly contrary to what she'd told the press when she'd resigned several years earlier. Then, she'd said she "still believe[d] in charters like Harlem Success because I've seen firsthand the amazing things they are doing for children," that she'd resigned because it "wasn't a sustainable life," that resigning had been "one of the hardest choices I've ever had to make" and that it had made her "feel, on a personal level, like a failure."[70] Jane had then become the media's "poster girl for charter school burnout"[71] and two *Times* articles had reported she'd resigned because she'd "burned out."[72] Incredibly, however, not only did Taylor report Jane's new story, which was flatly contrary to Jane's prior one, Taylor didn't even inform her readers about the inconsistency.

As for Jane's claim that we'd encouraged her to belittle children, precisely the opposite was true. In 2009, for example, one of our principals wrote to Jane that he'd felt she "needed to recognize [student] progress more" and, when she'd made progress, wrote that he was "happy to see how positive and reinforcing you are with your scholars." A year later, however, a student complained to me about Jane, as I wrote to her principal:

> [The student said] kids feel Jane is mean. Being strict and mean are different. Important that we watch for tone and positivity and nurturance. [I] know [you are] working on [it].

However, despite the principal's efforts to help Jane, I got complaints from parents that Jane was "too strict" and had to meet with them about this, which was quite unusual.

Another source of Taylor's was an assistant teacher I'll call Jill who said she'd "resigned . . . because she was so uncomfortable with the school's approach," which was like "witnessing child abuse." Again, this was demonstrably untrue. Jill's principal, a wonderful woman who'd been my own daughter's third-grade teacher, had seen problems with Jill's performance including "yelling at scholars," "raising [her] voice" when "frustrated," and "texting on phone while crossing the street with class," a serious safety issue. Jill responded that she didn't like "being judged in a negative way," "found it very difficult to sit through sessions of feedback," was unhappy that she was "not on the right path to becoming an LT [lead teacher]," and was upset because teaching was "the first thing I haven't been good at." She soon quit. In the email she sent explaining her decision, she didn't say one word about disagreeing with our educational approach, much less "witnessing child abuse." To the contrary, she said that her "heart was . . . invested" in "this job," but that she'd been "criticized . . . to a point where I knew I could no longer make it work here."

Again, I'm not faulting either Jill or Jane. It's often hard for people to see painful things objectively. It was Taylor's responsibility, however, to fact-check her sources. Rather than do so, she deliberately sandbagged us. She never told us what Jane was claiming so we couldn't respond to it by suggesting, for example, that Taylor check her own paper's news clippings. As for Jill, Taylor didn't tell us about her claims until 6:14 p.m. the night before her story appeared. We immediately protested to Taylor that she was "ambushing" us with "serious allegations" to which "we can't possibly respond this quickly," particularly since I'd never even heard of this employee, but Taylor denied our request that we be given time to look into Jill's claims. Like a detective who is so convinced a suspect is guilty that he fabricates evidence, Taylor didn't want to learn anything that might undermine the credibility of the sources who supported her version of the truth.

Taylor wrote many articles in addition to those I've described here. Incredibly, while she was the *Times*'s principal education reporter, 34 percent of the total words she wrote on education over the course of a year—sixteen articles in total and four of her seven longest ones—were devoted to negative coverage of Success and me personally. None of her other articles involved serious investigative reporting and several were puff pieces. She wrote approvingly, for example, of PS 191, a district school where only 12 percent of the students passed the state test in English and 11 percent in math. Taylor suggested, however, that the true measure of PS 191 was DOE's survey of the school's parents: "94 percent described themselves as either satisfied or highly satisfied with their child's education." Curiously, although DOE administered this same anonymous survey to parents at charter schools, Taylor didn't report on these results for Charlotte's school. They showed not only that 98 percent of our parents were satisfied but that 82 percent were *highly* satisfied (compared with only 42 percent of PS 191's parents) and that not *one single parent* in our entire school selected "teaching" as something that needed improvement.

As for the pesky little fact that PS 191 wasn't teaching its students to read, there was a simple explanation: "73% of students attending PS 191 qualified for free or reduced-price lunch" and "what the tests are really measuring . . . is the privilege of the children and the parents' educational level." In other words, parents who aren't well educated shouldn't expect their children to be taught to read well enough to pass the state English test. (Of course, we manage to do so with students who are poorer on average than those at PS 191—but silly me, I forgot that we fake our scores by secretly beaming 90 percent of our kids to the planet Krypton.)

Taylor also reported approvingly of DOE's intention to start "a gifted and talented program" at PS 191 to "attract wealthy parents to a mostly poor, nonwhite school," without noting that having separate classes for wealthy white students isn't the most authentic and beneficial form of integration. Nor did she include a single negative statement by an anonymous source or disgruntled former employee, although she found column space to note that PS 191's "parents and teachers described the atmosphere as friendly and collaborative."

While Taylor spent all of her time investigating Success, she ignored instances in which the district schools had knowingly tolerated far worse conduct than that in which Charlotte had engaged including:

A teacher at PS 194, who'd been reprimanded three times for corporal punishment but kept his job for years until he was finally arrested for throwing a seven-year-old special needs student across the hallway.

A teacher at PS 101 who wasn't fired despite the demand of parents and eleven complaints of verbal abuse and corporal punishment, six of which were substantiated.

A teacher at MS 172 who wasn't fired despite parent protests and the fact that he'd been caught "showing pornography to a fourteen-year-old student, buying him expensive gifts, sending him 513 text messages, and letting the student drive his car."

From the outset, it was obvious that the *Times* was taking a very different approach to reporting on Success than other schools. Taylor's first article on us asked parents of current and former Success students to contact the *Times,* a request the *Times* rarely makes. Their next article, which linked to this request, was listed on the *Times*' Internet home page for six months although the *Times* ordinarily rotates out its education articles every three days so new ones can be listed. The following article that Taylor wrote contained yet another request that parents contact the *Times.* While all of this was going on, we got reports for months on end that Taylor was doing endless investigative work such as calling through lists of former teachers.

I'm not saying that Success should be immune from scrutiny, but if a newspaper as powerful as the *Times* assigns a reporter to spend a year digging up every negative thing it can find about a school and encourages her to report what she finds as negatively as possible, the result is inevitably misleading. Readers naturally will think that there are a disproportionate number of problems at that school rather than a disproportionate amount of resources being spent on finding problems with that school.

Now I'm going to share with you some facts about Taylor and her editors that I fear may come across as an ad hominem attack but I hope you'll ultimately conclude isn't. Taylor graduated from a private school that offered sailing and golf and whose reputation, according to a book she edited, was that "every girl had a 'Saab, a shrink, and an eating disorder.'" Similarly, Taylor's immediate editor, Amy Virshup, graduated from a wealthy suburban school that had skiing, golf, and sailing teams and where less than 5 percent of the students qualify for free or reduced-priced lunch and only 7 percent are students of color. Finally, Virshup's boss, Wendell Jamieson, attended a private school where tuition currently runs $40,000 annually, although he ultimately graduated from a selective public high school.

There's nothing wrong with going to schools like these. Indeed, I sent my eldest son to a private high school. Neither is it surprising that the *Times* would tend to hire graduates of these schools since they are usually quite capable and well educated. However, when an institution is filled with reporters who have so little experience with the type of schools 99 percent of Americans attend, it can create a blind spot.

At Success, we aim to give our kids just as good an education as they'd get at a private or suburban public school, but we can't run our schools the same way. It's harder to maintain order when you have thirty kids in a class rather than fifteen and when many of your students have experienced emotional trauma. Moreover, while students at private schools learn to strive from their hyperachieving parents, many of our students have to be pushed to reach their potential.

We are nothing like Taylor's caricature of our schools. If we were, then parents wouldn't send their kids to our schools, particularly in neighborhoods like Cobble Hill and the Upper West Side where there are decent alternatives—but our schools are more strict than most district schools and I believe that Taylor and her editors were hostile to our approach because they just didn't understand the need for it given their backgrounds.

When Taylor was writing her first article on Success, I sensed that she didn't know much about public schools compared to other reporters with whom I'd dealt, so I asked her to spend a day at one of the district schools with which we were co-located so she'd understand what happens when a school serving disadvantaged children doesn't have high standards for student behavior and a rigorous approach to academics. Given that we'd acceded to Taylor's request to spend four full days in our schools unescorted, I thought asking her to spend one day in a district school was quite reasonable but she refused to seek permission to do so, claiming that she already knew what she'd find.

According to Taylor, her piece on Charlotte raised the question "How much do parents know of what goes on in their children's classrooms?" Of course, it was this attitude that led her to largely ignore what the parents she'd interviewed had told her. I thought, however, that her piece raised a different question: how a reporter with so little knowledge of urban public schools could be so confident that watching a seventy-second video and speaking to a few demonstrably unreliable former employees made her better qualified to know what Charlotte's classroom was really like than the parents of Charlotte's students who visited her class regularly, spoke with their children daily, and in most cases were actually far more knowledgeable than Taylor about public schools.

But frankly, I blame Taylor's editors more than Taylor herself. Taylor is an intelligent, hardworking reporter but, as Carl Bernstein once observed, "Reporters need good editors" and "this collaboration is what anchors the credibility of the press." Rather than guiding Taylor, however, her editors just egged her on. Jamieson publicly announced that "You can't have a bad day like that with a first-grader—I don't care," which struck me as an awfully emotional and unsympathetic view of the challenge of teaching. He and Virshup also claimed that they knew from the "body language" of the students in the seventy-second video that Charlotte acted like this frequently, evincing a confidence in their powers of observation rivaling that of Sherlock Holmes. Virshup also claimed she'd given us "every opportunity to . . . understand what our story was going to say ahead of time," which was patently false given how the *Times* had sandbagged us with Jill's and Jane's comments.

Unfortunately, the damage these stories did wasn't just to my feelings (although, yes, it does hurt to know that hundreds of thousands of people have read that I run schools that engaged in "child abuse"), but more important to our ability to educate our kids. After Taylor's stories came out, Andy Malone, now principal of our high school, emailed me about a teaching candidate who'd visited our school:

> She said, "I'm so relieved coming here, I read the article and I thought it had to be inaccurate, but now that I've seen the school I know that everyone is so loving here, the kids are so happy." Which was comforting but also infuriating. I'm just so angry for our kids and parents and faculty.

While candidates who visit our schools learn the truth about them, many potential applicants don't bother to apply in the first place. Frankly, it amazed me we got any applicants at all after Taylor's articles came out. If our schools were in fact anything like what she described, I wouldn't want to teach in them, much less devote my every waking hour to them and let them educate my own children.

But I feel most sorry for Charlotte and Candido. When their friends, parents, prospective employers, or romantic interests Google their names, the first page of results will consist largely of articles that wrongly condemn them for a single failure. They are fundamentally good and talented people who have devoted their lives to education and have suffered enormously and unfairly. I hope that through this book, more people will understand the truth about Charlotte and Candido, two passionate and talented educators who became casualties of the education wars.

43

THE EDUCATION OF EVA MOSKOWITZ

2003–2005

I was hopeful my committee's hearings would contribute to real changes in the teachers' union contract, which had expired in May 2003 and was now being renegotiated. Throughout 2003 and 2004, the city held firm, refusing to sign a contract that preserved "lockstep pay, seniority and life tenure," which, said Klein, were "handcuffs" that prevented him from properly managing the system. In June 2005, however, the UFT brought twenty thousand teachers to a rally at Madison Square Garden at which Weingarten demanded a new contract and Bloomberg's prospective Democratic opponents in the upcoming mayoral election spoke. The message was obvious: sign a new contract or we'll back your Democratic opponent. In October, the city capitulated, signing a new contract with none of the fundamental reforms sought by Klein.

This development accelerated a shift in my views on public education. I already supported charter schools, but I'd nonetheless held the conventional view that most public schools would and should be district run. I'd begun, however, to question that view. Every year, more children attended charter schools and you didn't have to be Einstein to see that there would come a day when most did if this trend continued. Maybe, I thought, this wouldn't be such a bad thing. Maybe a public school system consisting principally of charter schools would be an improvement.

This change of heart wasn't sudden. I didn't go to sleep one night believing in traditional public schools and wake up the next morning believing in charters. Rather, my views on school choice evolved gradually from profound skepticism, to open-mindedness, to cautious support, and were the product of decades of experience with the public schools as a student and then as an elected official.

At the very first school I attended, PS 36 in Harlem, I saw just how poorly some students were being educated. Through my work with Cambodian refugees in high school, I saw that good public education was largely reserved for those who could afford expensive housing. As a council member, I increasingly came to understand how the public school system's design contributed to segregation and inequality.

While it won't come as news to most readers of this book that schools in poor communities tend to be worse, understand that there is a difference between reading about this in the newspaper or a book and coming face-to-face with a mother who is desperate because she knows her son isn't learning anything at the failing school he is attending. Understand that there is a difference between knowing in the abstract that there are schools at which only 5 percent of the children are reading proficiently and actually visiting such a school and seeing hundreds of children who are just as precious to their parents as mine are to me but who you know won't have a fair chance in life because of the inadequate education they are receiving. Firsthand experiences like these cause you to reexamine your views carefully, to make absolutely certain they aren't based on faulty assumptions or prejudices or wishful thinking.

As a council member, I'd also become increasingly aware of the school system's dysfunction. In this book, I've recounted some of what I saw: textbooks that arrived halfway through the school year; construction mishaps; forcing prospective teachers to waste half a day getting fingerprinted. Know, however, that these are just a few select examples of a mountain of evidence that came to my attention from one hundred hearings, three hundred school visits, and thousands of parent complaints that came to me as chair of the Education Committee.

Moreover, even at their best, the district schools weren't innovative or well run, a point made by the late Albert Shanker, who was head of the American Federation of Teachers:

> [P]ublic education operates like a planned economy, a bureaucratic system in which everybody's role is spelled out in advance and there are few incentives for innovation and productivity. It's no surprise that our school system doesn't improve; it more resembles the communist economy than our own market economy.

While I was already convinced that the district schools weren't in good shape, preparing for the contract hearings was nonetheless an eye-opener for me. Interviewing principals, superintendents, and teachers helped me understand just how impossible it was for them to succeed given the labor contracts, and how job protections created a vicious cycle. Teachers felt they'd been dealt an impossible hand: their principal was incompetent or their students were already woefully behind or their textbooks hadn't arrived or all of the above. They didn't feel they should be held accountable for failing to do the impossible so they understandably wanted job protections. However, since these job protections made success even harder for principals to achieve who were already struggling with other aspects of the system's dysfunctionality, they too wanted job protections. Nobody wanted to be held accountable in a dysfunctional system, but the system couldn't be cured of its dysfunction until everyone was held accountable.

Some felt the problem was that the people entering the teaching profession tended to be weak, but I'd seen plenty of idealistic and intelligent teachers on my school visits. The system's dysfunction, however, took its toll on them. Some became so dispirited they left the profession or went to a suburban school; others burned out and became mediocre clock punchers; some heroically soldiered on, but even they rarely became the teachers they could have been.

Others claimed the solution was to increase education funds and reduce class size. There are limits, however, to how much we can afford to spend on education, and it's not clear it would make much of a difference anyway. Take PS 241, which is co-located with one of our schools. In the 2014–2015 school year, it had an average class size of just 12.7 students and spent $4,239,478 on one hundred kids,[73] $42,394 per student, but only two of those students passed the reading test that year.

In order to have any chance of fixing this system, I came to believe, we needed to radically change the labor contracts, which in turn required having elected officials who were willing to disagree with the UFT and stand up for children. I hoped to advance that goal by showing that even if you were independent of the UFT, you could survive politically. Obviously, that plan failed and the result was the opposite of what I'd hoped. Elected officials were more afraid of the UFT than ever and would tell Chancellor Klein, "I ain't gonna get Eva'd."

And I wasn't the only one the unions went after, as the *Times* reported:

> After trying to press [Gifford] Miller to cancel the hearings, Randi Weingarten, president of the teachers' union, attended the hearings with Brian M. McLaughlin, president of the New York City Central Labor Council, sending a message that the unions as a whole were displeased . . .
>
> "The old sense of discipline is gone, much to our disappointment," said a union official close to Ms. Weingarten. When she was stopped as she left City Hall last week and asked to discuss Mr. Miller, Ms. Weingarten frowned.

With Gifford now on the outs, he did terribly in his bid to become mayor and the UFT soon managed to restore the "old sense of discipline." Describing my hearings as a "biased," "McCarthy-like" "star chamber," Weingarten and McLaughlin made clear that the council should never again hold such hearings. The message was received. Council Member Robert Jackson, who became the next chair of the Education Committee, agreed that the council "should[n't] be diving into examining contracts."

I was also deeply troubled by the city's announcement in June 2005 that the number of students passing the state English test had increased by an astonishing 14 percent. I suspected the test had simply gotten easier because district schools throughout the state had achieved similar or even larger gains. My suspicions were confirmed by the results of the National Assessment of Educational Progress, a test that isn't changed to ensure accurate year to year comparisons and which showed no increase whatsoever in the number of students testing "proficient" and a miniscule increase in the number of students achieving "basic" literacy.* Yet at hearings I held, DOE claimed that the increase on the state tests was "not the result of differences in how the tests are constructed" but rather was due solely to "students' increased knowledge and skills." Not only did this fail the straight-face test, it fundamentally undermined the administration's case for reform. You can't simultaneously claim that you are handcuffed by the labor contracts and that you're hitting the ball out of the park. They are completely incompatible messages.

My point here isn't to criticize Bloomberg. He was quite savvy and had the best of intentions, so I presume that if he felt he had no choice but to play up the test scores and sign a bad contract with the UFT to win reelection, he was right. However, if even somebody as powerful and committed as Bloomberg couldn't make fundamental reforms, I didn't see how anybody else could. One columnist had observed, "If not now, when? If not Bloomberg, who?"[74] The answers to those questions, I reluctantly concluded, were "never" and "nobody."

My pessimism about reforming the district schools led me to take increasingly seriously the argument that we'd be better off if public education was principally provided by charter schools that were freed from the labor contracts, politics, and the stifling bureaucracy that plagued the district schools. I also wasn't persuaded by the arguments against charter schools: that they wouldn't be "accountable"

---

* https://nces.ed.gov/nationsreportcard/pdf/dst2007/2008455.pdf.

because they weren't government run, which struck me as weak given the government's abysmal record running district schools; and that parents couldn't be trusted to choose schools wisely, which was contrary to my experience that parents were in fact critical educational consumers and sent their children to weak schools not because they were blind to the flaws of these schools but rather because they lacked better alternatives.

The most baffling argument of all was that a district school system ensured equality since everybody was educated in the same system. In reality, the district school system was rife with inequality. I came to understand that the people who believed this argument weren't comparing charter schools to the public school system that actually existed but rather to their theoretical ideal. They imagined a system that was integrated and fair, in which every child got an adequate education, in which class size was small and every teacher was brilliant and nurturing. While it's admirable to aspire to these things, children attend school in the real world, and it's unwise to reject charter schools in favor of an ideal unless there's good reason to believe that the ideal is truly achievable. I'd come to believe it wasn't.

I didn't come to this conclusion lightly. After all, not only had I attended district schools, so had my parents, my brother, every one of my relatives who'd been born in this country, and my son. My grandmother and husband had even taught in the district schools. It felt almost disloyal to consider an alternative. Thus, even after my contract hearings in 2003, I continued to make fixing the district schools my top priority. I sought to address problems at particular schools that were brought to my attention and held hearings on a myriad of topics including teacher training, school safety, and science instruction, while handling other noneducational issues that arose including a dispute that was brewing about a sidewalk café application for a restaurant named Le Bilboquet.

What will the future hold for me? Part of me would love to be mayor for the simple reason that I love my city. Walt Whitman said, "There is no place like it, no place with an atom of its glory, pride, and exultancy. It lays its hand upon a man's bowels; he grows drunk with ecstasy; he grows young and full of glory, he feels that he can never die." John Lennon, who died a New Yorker, said he regretted not having been born one. Ed Koch, in his later years, said, "I wake up every morning and say to myself, 'Well, I'm still in New York. Thank you, God.'"

I love New Yorkers' "chutzpah," a Yiddish word for an attitude that is a cocktail of gall, audacity, brazen nerve, effrontery, guts, and arrogance. It is epitomized by a line in the film *Casablanca* that was penned by New York screenwriters Julius and Philip Epstein. When the Nazis ask Rick how he'd feel if they invaded his home town, he replies: "Well, there are certain sections of New York, Major, that I wouldn't advise you to try to invade."

I also feel a debt to New York City because it provided a refuge for my family when they fled from the Nazis, pogroms, and poverty. If America is the land of opportunity, New York is the city of opportunity and I'd like to make sure it continues to be.

I'm no longer sure, however, that I can best serve my city by becoming mayor. Perhaps I can do more for it by continuing to open more schools and improving the quality of education Success provides. When we began Success in 2006, 7 percent of the public school students in District 5 Central Harlem attended charters. Now, a decade later, 50 percent do, and it's having an enormous impact. In

2006, District 5 ranked 28th out of New York City's thirty-one school districts based upon the percentage of its elementary students, both district and charter, passing the state exams. Today, District 5 ranks 14th. This same pattern can be found throughout the city. Of the eight districts in which the highest percentage of students attend charter schools, all but one increased in the rankings over the last ten years, jumping eight spots on average. By contrast, not one of the districts that had a below average percentage of students in charters increased in the rankings and most fell. By increasing the number of charter schools across the city, we can dramatically increase educational opportunity.

Success now serves fifteen thousand students, making us equal in size to the entire public school system of a small city like Springfield, Illinois, but that's still less than 2 percent of New York City's public school population. We need to reach more students. We also need to improve the quality of the education we provide. Consider what is possible. James Mill, a nineteenth-century Scottish philosopher, attempted to find out how much a child could learn by having his own son undergo a rigorous course of study from a young age. His son began studying Greek at three and, having read many of its leading works by the age of eight, turned to Latin and algebra and thereafter to history, literature, science, and economics. By the age of twelve, this boy had an education far superior to that of most Ivy League graduates today. That boy, John Stuart Mill, went on to become a noted philosopher and economist. Can children learn as much today? And if not, how much can they learn? I don't know, but I do know that what we're achieving at Success is far short of what is possible, and that people will someday look at the education we offer today much as we now look at travel by horse and buggy.

But to get there, we need to make tremendous investments. Some people think teaching young children is simple because it's all "easy stuff," but while the content itself may be easy, figuring out how best to teach it isn't. When we think about teaching, we usually think about explaining. To teach a word, we explain its meaning; to teach a grammatical rule, we explain how it works. Children, however, have a tremendous capacity for intuitive learning. Eight-year-olds learn seven new words per day primarily just by hearing them used. Just by listening and reading, children learn complex rules of grammar they don't even consciously understand. Our entire approach to teaching may therefore be wrong. At a minimum, a tremendous investment needs to be made in creating an ideal curriculum from kindergarten through twelfth grade. How big an investment? Well, it's certainly more complicated than designing a new car and it costs around $6 billion. We've only scratched the surface of what is possible, which is why I find the work at Success so exciting and fulfilling.

This outcome was by no means inevitable. In 1991, I was miserable. I had no confidence in myself or any plan for my future. Perhaps I'd have been reasonably happy as a professor but it would not have given me the excitement and satisfaction I've experienced as a public servant and educator. Getting these opportunities wasn't the result of brilliant planning. On the contrary, as writing this book has made me profoundly aware, I got lucky. I don't mean that just in the philosophical sense of saying that everyone should appreciate and be grateful for the good things in their life. I mean it in the most literal sense: I, Eva Moskowitz, happened to get an astonishing number of lucky breaks in life, and I believe that anybody who gets as many lucky breaks as me should at least have the decency to own up to it.

To begin with, I'm lucky even to have been born. Had my mother not escaped the Holocaust, which took six million Jews, including my great-grandmother Raice Margulies, or had my mother died of typhus like many other passengers on the SS *Navemar,* I would not be here today.

I was lucky to have such wonderful parents who taught me to love learning, to work hard, to value family, and who gave me more than my share of chutzpah.

I was lucky that I saw an article about Gifford Miller running for the city council at just the moment when I had some time to volunteer on his campaign and that he turned out to be a candidate worth supporting. So much flowed from that one chance.

I was lucky that, although I lost my first election, my opponent soon resigned, giving me the perfect opportunity to win the seat. And I was lucky that Gifford then ran for speaker and won, which enabled me to become chair of the council's Education Committee.

I was lucky the *Times* didn't endorse me for borough president since losing that race led to my founding Success.

I was lucky that Joel Greenblatt and John Petry had a vision for a network of charter schools that would offer better opportunities for students across the city and that they gave me the opportunity to run them.

I was lucky to be given the chance to work with so many incredibly talented and idealistic educators, only a few of whom I can manage to mention in this book, but all of whom deserve credit for Success's achievements.

I was lucky that families gave us the opportunity to educate their children and that they have fought so hard to protect our schools.

I was lucky that Michael Bloomberg was elected mayor and that he appointed Joel Klein as chancellor. Without their support, Success Academies would be only a shadow of what it is today.

I was lucky Ed Cox was bold enough to lead the charge in granting Success three more charters when we didn't have any test results for our first school.

I was lucky that when we needed funding to grow, so many philanthropists gave generously including Daniel and Margaret Loeb, Julian Robertson, John Paulson, John Scully, Don Fisher, Eli Broad, the Walton family, and many others.

I was lucky that the editorial boards of the New York *Daily News* and the *New York Post* have been such unflagging advocates for charter schools and educational quality.

Most of all, I was lucky that I met Eric. He has been both the love of my life and my partner in much of what I've accomplished. Had he not seen my potential and encouraged me, I would never have entered politics, and nothing that you read in this book would have happened. While Eric has never wanted to share the limelight, the advice and assistance he has given me at every turn has been immeasurable.

And I am so grateful that I was able to conceive three healthy children when at one point it seemed doubtful that I'd bear even one. I'll never forget the moment that I held Eric's hand tight, mourning my lost child, only to see the faint glimmer of Culver's heartbeat on the sonogram screen.

What's next? Who knows. But one thing I've learned is that chance events largely determine your path. When Napoleon's aides recommended that an officer be promoted to general, Napoleon would respond: "I just want to know one thing: is he lucky?"

Yes, I am.

45

EXTRA CREDIT

Well, dear reader, we've reached the end of the road; that's my life story or at least as much of it as I've lived so far, but being an educator, I can't resist giving you a chance to earn extra credit. Your mission, should you choose to accept it, is to open up your mind to new ideas about education.

Success requires the will to question one's intuitions. To take an example from another field, India's government buys billions of dollars of food to address the problem of stunted growth among its country's children. Scientists, however, have observed something puzzling: stunting is common even in families that can afford ample food. The principal culprit, it turns out, is sanitation: infants are exposed to so much human waste that their bodies are constantly fighting disease which, by the age of two, results in irreversible stunting. The seemingly obvious explanation for stunting, that it was due to malnutrition, turned out to be wrong.

Human intuition isn't always a good guide to reality. Who would think that gravity could bend light or that a man traveling in a rocket ship at nearly the speed of light could return to earth to find himself ten years younger than his twin? But, as Albert Einstein taught us, these things are true. It's easy to laugh at the fact that people used to think the sun traveled around the earth or that you could cure people by bleeding them; what's much harder is to recognize that many of the things you believe now will seem just as foolish to people in the future.

Henry Fielding once observed:

> There are a set of religious, or rather moral writers, who teach that virtue is the certain road to happiness, and vice to misery, in this world. A very wholesome and comfortable doctrine, and to which we have but one objection, namely, that it is not true.

This tendency to imagine that the world is the way we'd like it to be, to look at it with rose-colored glasses, is commonplace in education. For example, we often idealize children. As Freud observed, human beings didn't evolve to live in civilized society. Neither did they evolve to attend school. If you start with the idea that children will naturally behave the way you want them to, that it's just like planting a seed and watching it grow, you will be disappointed and less successful at teaching them. You are more likely to succeed if you accept that schooling often requires getting children to act contrary to their natural inclinations.

Finding out what truly works in education takes real commitment, courage, and reflectiveness. You must be willing to reexamine your beliefs every day. Because we do this at Success, many people find it hard to pigeonhole us: we look like traditionalists if you focus on our approach to discipline, but like progressives if you focus on our approach to instruction. Which are we really? Neither, because what we do isn't about who we are, but what the kids need.

I've tried throughout this book to give you a sense of Success's educational philosophy and practices so I won't repeat all of them here, but I will share with you a few overarching themes and ideas. If you want to see what these principles look like in practice, go visit our virtual schools on the Success Academy Web site. Also, take a look at our Education Institute Library, which is also on our website. It reflects the efforts of our chief academic officer, Michele Caracappa, and her team to share with educators around the country the best practices we've developed over the last decade.

Here are our core educational values and practices:

Education Is for Kids, but About the Adults. Schools often focus on the changes children need to make, but it's the adults who must change first. If a principal goes into a classroom and sees that a couple of students in the back are fooling around and the teacher isn't noticing, the solution isn't to talk to the students, but to talk to the teacher about improving her classroom management. It is adult mastery of content and preparation, the caliber of adult questioning and facilitation, and the quality of the demands adults place on kids' thinking that determine a school's success. Many schools blame the students when they don't learn but it's the adults who are responsible for ensuring that students learn.

Content Matters. Curriculum in many schools has been dumbed down and is often incoherent. Worksheets involving repetitive low-level thinking abound. Schools need to invest heavily in selecting and designing high-quality content and curriculum. This means choosing excellent books but it also means investing in lesson planning. As Tony Lombardi said to my committee, "planning is the heartbeat of teaching."

Management Is Key. Excellence and high academic results don't just happen. You can't simply let a thousand flowers bloom. Educational leaders must be intentional and deliberate. They must manage their educational communities.

Young Children Need to Be Molded. Some educators think that schools can just tap into a child's natural curiosity and desire to learn. While schools should nurture those instincts, that alone isn't sufficient. Most children aren't naturally studious, so you need to use every tool in your chest to help them make the right choices and conduct themselves in a productive way. Developing children morally and intellectually must be an intentional process that the adults invest time and thought into; it takes enormous work and energy to help students become their best selves.

Suspensions Are a Useful Disciplinary Tool. A suspension is really just the equivalent of what at home is called a timeout. Being kept out of school for a day or two for breaking the rules communicates to a child that following the rules is a condition of being in school, a concept that both parents and students can readily understand. Suspensions also communicate to other members of the community that their safety is really valued and that our commitment to community values and norms is real.

Schools Need to Do a Better Job of Providing Students with a Safe and Productive Learning Environment. Most parents agree with this, which is one of the reasons we had 17,000 applicants apply for 3,000 spots at ours schools in 2017. While some parents complain when their own children are disciplined, these parents rarely remove their children from our school because it's most important to them that their own children are safe and aren't bullied or distracted from learning by other misbehaving children.

Children Need to Struggle. Many educators think their job is to cut up intellectual work into bite-size pieces. This approach minimizes intellectual challenge. Learning is like weight lifting: being pushed makes you stronger. In the words of our senior manager of chess, Sean O'Hanlon, it's supposed to be hard.

School Should Also Be Joyful. Some people think that a strict school can't be joyful. Nonsense. Just as in the real world, there is a time for work and a time for play. Not allowing kids to make wisecracks and fool around in the middle of math class doesn't mean students shouldn't have fun at school. It's a question of time and place. At Success, we have faculty vs student dodgeball games, watch slapstick films, hold contests at which students throw pies at their principal, go to the circus, and bring in jugglers and magicians. We also have an extremely robust student electives program since we believe that schools should help students discover their passion for art, dance, sports, theater, computer coding, debate, and music.

Small Class Size Isn't as Important as Many Educators Think. Everything else being equal, smaller class size is better, but everything else isn't equal. Having small class sizes is very expensive. Cut class size in half and you need twice as many teachers, which means virtually doubling a school's costs. That's why private schools with small class sizes often have a tuition of $50,000. New York City's district schools spend so much on teachers that they have little money left over for other things. They spend an average of less than 3 percent of their budget on the following: instructional supplies and equipment (1 percent), textbooks (0.6 percent), library books and librarians (0.5 percent), and computer support (0.5 percent). At Success every classroom has a SMART Board, a modern blackboard that is actually a gigantic touch-screen monitor; every teacher has a laptop, every conceivable classroom supply, and access to a catalog of lesson plans and videotaped lessons; and every student gets a laptop computer starting in fourth grade. We also pay our teachers more than the district schools do and give them far more professional development. We can do all this strictly on government funding that is less than the district schools by having larger class sizes.

Educators Should Be Promoted to Leadership Positions Based on Performance. While district school principals must have a minimum of seven years of teaching experience and a principal's license, none of Success's principals are licensed and few have this much teaching experience. Licensing is unnecessary because a school is in a better position than regulators to determine whether an educator is capable of becoming a principal. As for experience, the amount an educator needs before becoming a principal varies greatly: some require only a couple of years while others will never be ready to lead no matter how much experience they get. The key is investing in on-the-job training for principals to help them become great leaders and managers.

A School's Teachers Need to Function as a Team. Some teachers who join Success disagree with our philosophy and start marching to the beat of their own drum. The problem with this is not that we are necessarily right and they are wrong; reasonable minds can differ about how best to educate kids. The problem is that a school's educators need to function as a team, which means working from the same playbook. If a teacher can't in good conscience do that, she should get a job at a school whose philosophy she does support.

The same is true of parents. Some parents don't agree with our approach to education. That's fine. Different families have different values and want different things for their children. Parents who don't like Success should find a school they do like. For somebody to enroll their child at Success and insist we change our model is like a person walking into a pizzeria and demanding sushi. If you want sushi, go to a sushi restaurant!

Schools, however, have an obligation to be transparent with parents and teachers about their design, philosophy, and practices so that parents and teachers know what they are signing up for. When I speak at enrollment meetings for parents who are considering sending their children to Success, I'm very blunt with them about the features of our school they may not like. For example, a parent who feels strongly that elementary school children shouldn't have homework won't like our schools.

Teachers Should Have Lots of In-School Professional Development. Teachers at Success receive the equivalent of thirteen weeks of professional development per year compared to a few days in district schools. It's how we ensure high-quality teaching. Off the shelf professional development rarely works because teachers need to learn the playbook for their own school. Schools are also better situated to develop effective training for their own teachers because they can see the results. A school that outsources its teacher training is like a coach who doesn't run his own practices.

Principals Should Be Focusing Most of Their Efforts on Improving Teachers. Principals should regularly provide professional development, lead planning sessions, and make frequent brief classroom observations followed by immediate practical suggestions for improvement. For principals to focus on improving instruction, they need to be relieved of other responsibilities. At Success, we have business operations managers who manage many of the schools' critical operations including arrival, lunch, dismissal, procurement, facilities maintenance, and form collection. We also have educational managers who help students get special education services and provide data and analyses for principals so they can make the key strategic decisions that will improve teacher practice and student outcomes.

Hustle. Most teachers in America could dramatically improve their teaching if they just made every second count. In Michael Bloomberg's autobiography, he mentions that he found people were wasting a lot of time in meetings so he removed the chairs to change their mind-set about how long the meeting was expected to last. Speed and urgency matter.

Rigor. Teachers should pitch instruction at a high level. Think of kids at the dinner table listening to their parents having a political conversation: the children may not understand everything but they will still benefit from it and they will become familiar with the concepts over time. Similarly, not every kid has to understand every word or every idea in a lesson. It's better to aim a lesson too high than too low.

Going Beyond Z. Running good schools is a lot about simple hard work: doing the next step, doing the extra credit—just as you have by reading this chapter.

ACKNOWLEDGMENTS

I'm thankful to the many people who read and commented on this book and who have helped Success Academy thrive.

Success simply wouldn't exist without Joel Greenblatt and John Petry who started it and continue to provide great financial support and wise counsel to this day. It would not have thrived without Daniel and Margaret Loeb, who provided support at a critical moment in our history. Their first contribution allowed us to open up three new schools in Brooklyn and they have continued to be generous supporters. In 2013, Daniel took over as chair of our Network board, an enormous time commitment. Since then, he has been an incredible thought partner, contributing toward our mission and design in countless ways. He has brought innumerable politicians, journalists, and civic and business leaders to visit our schools, which has provided enriching learning experiences for our scholars and the visitors alike. Wielding his influence in countless ways, Daniel is really our Chief Advocacy Officer for children.

Sam Cole leads the board for our schools (a separate entity from the Network). It is the most time consuming of all of the board positions and his willingness to find the time for it shows his deep commitment to our schools and the children they serve.

The late Don Fisher gave us our first substantial outside philanthropic contribution, and the Fisher family, with Doris and John now leading the way, have continued to support us at every turn. Eli and Edye Broad were also early and generous supporters. Their leadership in the arts, sciences, and education reform puts them in the pantheon of American philanthropists. John and Regina Scully joined our board, gave us our first eight-figure gift to jump-start our scaling mission, and have hosted numerous events for us in their beautiful home in San Francisco.

Bruce and Suzie Kovner have become true mainstays, starting with early support from the Kovner Foundation that has grown into transformative funding for the Success Academy Education Institute, Suzie's service as a member of the board, and much more. Roger Hertog supported my first run for office, has advised me ever since, and, along with his wife Susan, has provided generous support for the Education Institute.

John and Jenny Paulson have funded our second high school, the renovation of the historic Bronx Borough Courthouse where it will be located, and two middle schools. John has also provided an invaluable service by using his position as an important civic leader to advocate on behalf of our schools. Board member Tali Farhadian Weinstein and her husband, Boaz Weinstein, have started two elementary schools, and they have been deeply engaged in providing wonderful learning experiences for our scholars and faculty.

In 2013, Julian Robertson agreed to fund the Success Academy Education Institute, which is providing a digital platform for all of our internal training needs as well as the external vehicle for sharing with other schools the curricula, pedagogy, and training we have developed over the past decade. Julian's support enabled the complicated task of codifying, streamlining, organizing, and digitizing all of our curricular and training materials. He subsequently gave us a $25 million gift, by far the largest we've ever received. I've heard Julian speak movingly about how his philanthropic endeavors stem from the commitment to community that he learned as a child from his parents and I hope that, just as he learned from their example, others will learn from his.

My husband, Eric, and our three children—Culver, Dillon, and Hannah—have not only sustained me personally but have each made innumerable contributions to Success Academies. My love for them knows no bounds. My parents, Anita and Marty, taught me the value of education and what it means to stand up for an ethical society.

My mother-in-law, Alex Grannis, has supported my career at every turn. I'm especially grateful to Jocelyn Galvez, the executive assistant everyone wants, who makes all possible with my crazy life.

Finally, I am profoundly grateful to everyone at Success who works so hard day in and day out. This includes not only our educators but also our students and their parents. Their commitment to educational excellence is what keeps me going.

NOTES

As a trained historian, it was important to me that this book be completely accurate, so it was meticulously researched. Tens of thousands of emails, hundreds of articles, and numerous documents obtained through freedom of information law requests were reviewed. In addition, many people were interviewed.

For the sake of brevity, I have deleted portions of my quotes and those of people associated with Success Academy or my family without indicating ellipses because I believed they would be distracting and of little utility. Examples of this include my emails as well as those of Khari Shabazz and Andrew Malone, the teacher entries quoted in the chapter concerning Kate Taylor's coverage, and my grandfather Chaim's poems and diary entries. I did not take this liberty with respect to anybody not associated with Success Academy or me personally.

For fact-checking purposes, drafts of this book had nearly 500 endnotes, many to internal Success emails, but I have deleted nearly all of them in the published text because they served little purpose. Where I quote a newspaper, I've generally identified the newspaper in the text so it can be located with an Internet search. The endnotes below reflect only those instances in which I believed it would be useful to the reader to have a citation. Throughout this book, I cite information regarding test scores, class size, demographics, and budgets for both Success's schools and district schools. That information is contained in the New York State Education Department's schools report cards and in the New York City Department of Education's progress reports, school surveys, school-based expenditure reports, school quality snapshots, registers, and average class-size reports. All of the foregoing reports are readily available on the SED and DOE websites simply by searching for the school in question. Information regarding school utilization can be obtained by searching on the website of the School Construction Authority (www.nycsca.org) for the enrollment, capacity, and utilization reports, also known as the "Blue Book."

1. Monty Phan, "Madison Avenue," *Newsday*, September 27, 2004, accessed May 31, 2017, http://www.newsday.com/business/technology/madison-avenue-1.692050.

2. David M. Herszenhorn, "Mayor's Goal Is 'Thin' Pact With Teachers," *New York Times*, February 05, 2004, accessed May 31, 2017, http://www.nytimes.com/2004/02/06/nyregion/mayor-s-goal-is-thin-pact-with-teachers.html.

3. Ryan Sager, "Mayor On the Edge of a Sellout—Letting Teachers Union Off The Hook," *New York Post*, October 20, 2004, accessed May 31, 2017, http://nypost.com/2004/10/20/mayor-on-the-edge-of-a-sellout-letting-teachers-union-off-the-hook/.

4. Cassi Feldman, "Working Families Fracas: Dems Blast Party for Meddling," *City Limits*, September 12, 2005, http://citylimits.org/2005/09/12/working-families-fracas-dems-blast-party-for-meddling/.

5. Wayne Barrett, "Bloomberg and the Teachers' Union," *Village Voice*, May 13, 2009, accessed May 31, 2017, https://www.villagevoice.com/2009/05/13/bloomberg-and-the-teachers-union/.

6. "Randi Weingarten," AFT Facts, accessed May 31, 2017, https://www.aftfacts.com/randi-weingarten/.

7. Hannah Arendt, *The Portable Hannah Arendt* (New York: Penguin Books, 2003), 321.

8. Elissa Gootman, "Moskowitz, Critic of Education Department and Union, Will Head a Charter School," *New York Times*, December 10, 2005, accessed May 31, 2017, http://www.nytimes.com/2005/12/11/nyregion/11eva.html.

9. Ibid.

10. "Children's Access to Care," American Federation of Teachers, accessed May 31, 2017, https://www.aft.org/childrens-health-safety-and-well-being/childrens-access-care.

11. "United Federation of Teachers Comparison of 2002 & 2009 Governance Laws," United Federation of Teachers, accessed June 14, 2017, https://www.uft.org/files/attachments/uft-governance-101.pdf.

12. Maura Walz, "Harlem Success students welcomed back with a protest." *Chalkbeat*, September 9, 2009. Accessed June 14, 2017. http://www.chalkbeat.org/posts/ny/2009/09/09/harlem-success-students-welcomed-back-with-a-protest/.

13. "PS123 & Harlem Protest HSA Charter Invasion," YouTube, posted by "Angel Gonzalez," September 13, 2009, accessed May 31, 2017, https://www.youtube.com/watch?v=H42lPSqjiHg.

14. "Neighborhood News," *DNAinfo New York*, accessed May 31, 2017, https://www.dnainfo.com/new-york/places/upper-west-success-academy.

15. Thomas K. Duane, "Statement By New York State Senator Thomas K. Duane Before the New York City Department of Education, the District 3 Community Education Council And School Leadership Teams Regarding the Proposed Co-location of Upper West Success Charter School With Existing Schools in the Brandeis Educational Campus," January 25, 2011, accessed May 31, 2017, https://www.nysenate.gov/sites/default/files/articles/attachments/DuaneTestimony%20on%20Upper%20West%20Success.pdf.

16. *Unequal Shares: The Surprising Facts about Charter Schools and Overcrowding*, Report, New York Charter School Center, October 2011, accessed June 15, 2017, http://www.nyccharterschools.org/sites/default/files/resources/unequal_shares.pdf.

17. Michael Barbaro, "Criticized as Too Sedate, Public Advocate's Office Intends to Get Louder," *New York Times*, January 3, 2010, accessed May 31, 2017, http://www.nytimes.com/2010/01/04/nyregion/04advocate.html.

18. Wayne Barrett, "Barrett: Bill DeBlasio, Public Advocate or Teachers Union Patsy? (UPDATED)." *Village Voice*, January 12, 2010, accessed June 5, 2017, http://blogs.villagevoice.com/runninscared/2010/01/barrett_37.php.

19. Tom Topousis, "Council Catfight: The Contest For the Upper East Side's City Council Seat Is a Vicious Battle Royale," *New York Post*, October 13, 1999, accessed May 31, 2017, http://nypost.com/1999/10/13/council-catfightthe-contest-for-the-upper-east-sides-city-council-seat-is-a-vicious-battle-royale/.

20. Ibid.

21. "District 3 Press Conf Part 1," YouTube, posted by "GEMNYC1," October 21, 2010, accessed May 31, 2017, https://www.youtube.com/watch?v=v-2TYzOCrQs.

22. Conor Skelding, "Female officials on Moskowitz, Palin, age-reporting," Politico PRO, March 12, 2014, accessed May 31, 2017, http://www.capitalnewyork.com/article/city-hall/2014/03/8541779/female-officials-moskowitz-palin-age-reporting.

23. https://www.washingtonpost.com/news/education/wp/2016/05/17/on-the-anniversary-of-brown-v-board-new-evidence-that-u-s-schools-are-resegregating/

24. Nikhita Venugopal, "De Blasio Demands Investigation of Cobble Hill Charter School," *DNAinfo New York*, April 24, 2013, accessed May 31, 2017, http://www.dnainfo.com/new-york/20130424/cobble-hill/de-blasio-demands-investigation-of-cobble-hill-charter-school.

25. Ibid.

26. "De Blasio takes on city's treatment of Moskowitz charter schools," *Chalkbeat*, April 22, 2013, accessed May 31, 2017, http://ny.chalkbeat.org/2013/04/22/de-blasio-takes-on-citys-treatment-of-moskowitz-charter-schools/#.VHYJM4vay3w.

27. Bernard P. Orland, "PCB Wipe Sampling Report," May 8, 2013, accessed May 31, 2017, http://schools.nyc.gov/NR/rdonlyres/7F169C51-448A-43CB-8002-AE2669D51A8E/144293/M1232020PCB.pdf.

28. The Editors, "Editorial: Pandering to the UFT," *Observer*, May 14, 2013, accessed May 31, 2017, http://observer.com/2013/05/editorial-pandering-to-the-uft/.

29. Michael Powell, "Moskowitz from:powellnyt - Twitter Search," Twitter, February 22, 2017, accessed May 31, 2017, https://twitter.com/powellnyt/status/339483686875578370.

30. "In Debate Debut, Anthony Weiner Talks Education And Puts Gov. Cuomo On Notice," *NY Daily News*, May 28, 2013, accessed May 31, 2017, http://www.nydailynews.com

/blogs/dailypolitics/debate-debut-anthony-weiner-talks-education-puts-gov-cuomo-no tice-blog-entry-1.1694793.

31. Jim Epstein, "Sick: NYC's Bill de Blasio Puts Politics Before Poor Kids," Reason.com, March 9, 2014, accessed May 31, 2017, http://reason.com/blog/2014/03/09/reason-tv-re play-sick-nycs-bill-de-blasi.

32. Helen Zelon and Leah Robinson, "The 2013 Primary Candidates on Education," *City Limits*, October 27, 2014, accessed May 31, 2017, http://citylimits.org/2013/09/05/the -2013-primary-candidates-on-education/.

33. John Toscano, "Labor Council Names Committee To Deal With McLaughlin Probe," *Queens Gazette*, accessed May 31, 2017, http://www.qgazette.com/news/2006-03-08 /features/024.html; Juan Gonzalez, "Labor Has Plan to Handle Probe. Juan Gonzalez Exclusive on Bid-rig Investigation," *NY Daily News*, March 7, 2006, accessed May 31, 2017, http://www.nydailynews.com/archives/news/labor-plan-handle-probe-juan-gonzalez -exclusive-bid-rig-investigation-article-1.598212.

34. Steven Greenhouse, "Mayor Knew Of Investigation Even as Unions Backed Him," *New York Times*, March 4, 2006, accessed May 31, 2017, http://query.nytimes.com/gst /fullpage.html?res=9402E0DA1431F937A35750C0A9609C8B63.

35. Sarina Trangle, "PS/IS 49 to lose principal," *TimesLedger*, February 6, 2014, accessed May 31, 2017, http://www.timesledger.com/stories/2014/6/lombardi_tl_2014_02_07_q.htm.

36. Geoff Decker, "Sea of parents and advocates take to streets for charter schools," *Chalkbeat*, April 12, 2016, accessed May 31, 2017, http://ny.chalkbeat.org/2013/10/08/sea -of-parents-and-advocates-take-to-streets-for-charter-schools/#.VV8fLfmrSso.

37. Post Editorial Board, "Parents' historic march for charter schools," *New York Post*, October 6, 2013, accessed May 31, 2017, http://nypost.com/2013/10/06/parents-historic -march-for-charter-schools/.

38. Ben Chapman, "Charter school rally sends message to Bill de Blasio, Joe Lhota," *NY Daily News*, October 8, 2013, accessed May 31, 2017, http://www.nydailynews.com/new -york/education/charter-school-rally-sends-message-de-balsio-lhota-article-1.1479884.

39. "Second Revised Building Utilization Plan," Schools.nyc.gov, accessed May 31, 2017, http://schools.nyc.gov/NR/rdonlyres/7AC20E2C-5933-4F54-806B-DA19BE3A9C55 /149454/SAHarlem4inM149andM207BUP_vFINAL1.pdf; "Educational Impact Statement: The Proposed Co-location of Grades Five through Eight of Success Academy Charter School—Harlem 4 (84M386) with Existing Schools P.S. 149 Sojourner Truth (03M149), P.S. M811 Mickey Mantle School (75M811), and Grades Kindergarten through Four of Success Academy Charter School—Harlem 1 (84M351) in Tandem Buildings M149 and M207 Beginning in the 2014–2015 School Year," NYC Department of Education, August 30, 2013, accessed May 31, 2017, http://schools.nyc.gov/NR/rdonlyres/7AC20E2C -5933-4F54-806B-DA19BE3A9C55/149453/SAHarlem4inM149andM207EIS_vFINAL1.pdf.

40. MYFOX New York Staff. "Schools Chancellor Carmen Farina on rough first 2 months." Internet Archive, March 7, 2014, accessed June 5, 2017, https://web.archive .org/web/20140311011757/http://www.myfoxny.com/story/24913914/schools-chancellor-carmen-farina-on-rough-first-2-months.

41. Diane Ravitch, "New York Schools: The Roar of the Charters," *New York Review of Books*, accessed May 31, 2017, http://www.nybooks.com/blogs/nyrblog/2014/mar/27 /new-york-charters-against-deblasio/.

42. Valerie Strauss, "The Big Losers in NYD Charter Fight: Students With Disabilities," *Washington Post*, April 14, 2014, accessed May 31, 2017, http://www.washingtonpost

.com/blogs/answer-sheet/wp/2014/04/14/the-big-losers-in-nyc-charter-fight-students-with-disabilities/.

43. Mara Gay, "New York Charts Bold Course for Schools," *Wall Street Journal*, March 31, 2014, accessed May 31, 2017, http://online.wsj.com/news/articles/SB10001424052702304157204579473883631027514?KEYWORDS=school&mg=reno64-wsj.

44. Leslie Brody, "New York City Charter Move Proves Pricey," *Wall Street Journal*, May 28, 2014, accessed May 31, 2017, http://www.wsj.com/articles/new-york-city-charter-move-proves-pricey-1401326471.

45. Diane Ravitch, "New York Schools: The Roar of the Charters," *New York Review of Books*, March 27, 2014, accessed May 31, 2017, http://www.nybooks.com/blogs/nyrblog/2014/mar/27/new-york-charters-against-deblasio/.

46. Lizzy Ratner, "Taking on Unions, And Paying a Price," *Observer*, December 8, 2003, accessed June 15, 2017, http://observer.com/2003/12/taking-on-unions-and-paying-a-price/.

47. "We're the teechers round hear." *Economist*, November 20, 2003, accessed June 5, 2017, http://www.economist.com/node/2235294.

48. David Saltonstall, "It's a Pointed Lesson In Political, School Ties," *NY Daily News*, November 14, 2003, accessed May 31, 2017, http://www.nydailynews.com/archives/news/pointed-lesson-political-school-ties-article-1.514168.

49. Lizzy Ratner, "Taking on Unions, And Paying a Price," *Observer*, December 7, 2003, accessed May 31, 2017, http://observer.com/2003/12/taking-on-unions-and-paying-a-price/.

50. Ibid.

51. Kate Taylor, "New York City Teachers' Union Is Closing Portion of Its Brooklyn Charter School," *New York Times*, February 27, 2015, accessed May 31, 2017, http://www.nytimes.com/2015/02/28/nyregion/new-york-city-teachers-union-is-closing-portion-of-its-brooklyn-charter-school.html.

52. Post Staff Report, "The UFT proves a point," *New York Post*, October 11, 2012, accessed June 5, 2017, http://nypost.com/2012/10/11/the-uft-proves-a-point/.

53. "Opened to prove a point, UFT's charter school could be closed," *Chalkbeat*, October 9, 2012, accessed May 31, 2017, http://ny.chalkbeat.org/2012/10/09/opened-to-prove-a-point-uft-charter-school-could-be-closed/#.VnH5I7grLuQ.

54. Danny Wilcox and Michelle Bodden-White, "UFT Charter School 2009-10 Accountability Plan Progress Report," http://67.225.207.84/files/6114/7819/5112/Accountability_Plan_2009-2010.pdf.

55. United States, New York State Education Department, University of the State of New York, The New York State School Report Card: Accountability and Overview Report 2009-2010, February 5, 2011, accessed June 5, 2017, https://web.archive.org/web/20151103144844/https://reportcards.nysed.gov/files/2009-10/AOR-2010-331900860891.pdf.

56. Elissa Gootman, "Moskowitz, Critic of Education Department and Union, Will Head a Charter School," *New York Times*, December 10, 2005, accessed May 31, 2017, http://www.nytimes.com/2005/12/11/nyregion/11eva.html.

57. Eva Moskowitz, "De Blasio Allies Counterattack on Moskowitz—the Real Civil Rights Issue of the Time," FUTURE 231, accessed May 31, 2017, http://future231.blogspot.com/2014/03/de-blasio-allies-counterattack-on.html.

58. "Success CEO Eva Moskowitz builds suspense for Thursday announcement of 'political plans'," *Chalkbeat*, April 12, 2016, accessed May 31, 2017, http://ny.chalkbeat

.org/2015/10/07/success-ceo-eva-moskowitz-builds-suspense-for-thursday-announcement-of-political-plans/#.VoAtEMYrLnA.

59. Jillian Jorgensen, "Eva Moskowitz Is Going to Say Something Tomorrow," *Observer*, October 8, 2015, accessed May 31, 2017, http://observer.com/2015/10/eva-moskowitz-is-going-to-say-something-tomorrow/.

60. Beth Fertig, "Charter Leader Moskowitz Not Running for Mayor After All," WNYC, October 8, 2015, accessed May 31, 2017, http://www.wnyc.org/story/charter-leader-moskowitz-not-running-mayor-after-all/.

61. United States, New York City Independent Budget Office, "Comparing Student Attrition Rates at Charter Schools and Nearby Traditional Public Schools," January 2015, accessed June 5, 2017, https://web.archive.org/web/20160309171319/www.ibo.nyc.ny.us/iboreports/2015schoolattrition.pdf; Beth Fertig, "Harlem Schools See High Student Turnover." WNYC, October 2, 2012, accessed June 5, 2017, http://www.wnyc.org/story/302691-harlem-schools-see-high-student-turnover/; Beth Fertig, "Top 10: Charters with Highest Attrition Rates," WNYC, October 4, 2012, accessed June 5, 2017, www.wnyc.org/story/302728-top-ten-charters-with-high-attrition-rates/.

62. "NYC Top Charter Attrition Rates 2010-11," Google Sheets, accessed May 31, 2017, https://docs.google.com/spreadsheets/d/1e48Jl7bNHk06-6AQb4pt3yiTwfQvI3a1e_Df1nRaveA/edit?pli=1#gid=1.

63. "Comparing Student Attrition Rates at Charter Schools and Nearby Traditional Public Schools," accessed May 31, 2017, http://www.ibo.nyc.ny.us/iboreports/2015schoolattrition.pdf.

64. John Merrow, "Eva's Offensive," *The Merrow Report*, March 20, 2016, accessed May 31, 2017, https://themerrowreport.com/2016/03/18/evas-offensive/.

65. Rishawn Biddle, "Moskowitz, Merrow, and Over-Suspensions," Frontline, "*Dropout Nation*," October 21, 2015, http://dropoutnation.net/2015/10/20/moskowitz-merrow-and-over-suspensions.

66. Alexander Russo, "Some Questions For John Merrow, PBS, & Eva Moskowitz*," *Washington Monthly*, October 19, 2015, http://www.washingtonmonthly.com/the-grade/2015/10/some_questions_for_john_merrow058190.php.

67. "Why WNYC's District-Charter Attrition Comparison Has Gotten So Little Love," *Washington Monthly*, May 10, 2016, accessed May 31, 2017, http://washingtonmonthly.com/2016/03/24/why-wnycs-district-charter-attrition-comparison-has-gotten-so-little-love/.

68. Beth Fertig and Jenny Ye. "NYC Charters Retain Students Better Than Traditional Schools." March 15, 2016, accessed June 5, 2017, http://www.wnyc.org/story/nyc-charter-school-attrition-rates/.

69. "Why WNYC's District-Charter Attrition Comparison Has Gotten So Little Love," *Washington Monthly*, May 10, 2016, accessed May 31, 2017, http://washingtonmonthly.com/2016/03/24/why-wnycs-district-charter-attrition-comparison-has-gotten-so-little-love/.

70. Alexander Russo, "Meet the Poster Child for Charter School Burnout," *Huffington Post*, September 7, 2011, accessed May 31, 2017, http://www.huffingtonpost.com/alexander-russo/post_2370_b_950392.html.

71. Ibid.

72. Michael Winerip, "Teachers Get Little Say in a Book About Them," *New York Times*, August 28, 2011, accessed May 31, 2017, http://www.nytimes.com/2011/08/29/education/29winerip.html; Joe Nocera, "Teaching With the Enemy," *New York Times*, November 7, 2011, accessed May 31, 2017, http://www.nytimes.com/2011/11/08/opinion/teaching-with-the-enemy.html;

73. "FY 2015 Summary Reports," August 3, 2016, https://www.nycenet.edu/offices/d_chanc_oper/budget/exp01/y2014_2015/pdf/03.pdf.

74. Ryan Sager, "Mayor On the Edge of a Sellout—Letting Teachers Union Off the Hook," *New York Post*, October 20, 2004, accessed May 31, 2017, http://nypost.com/2004/10/20/mayor-on-the-edge-of-a-sellout-letting-teachers-union-off-the-hook/.

INDEX

ABOUT THE AUTHOR

EVA MOSKOWITZ is the founder and CEO of Success Academy Charter Schools and a former New York City council member. She has a PhD in American history from Johns Hopkins University and a BA from the University of Pennsylvania, and is a graduate of Stuyvesant High School. She lives in Harlem with her husband and their three children.